Intermediate Set Theory

Intermediate Set Theory

F. R. Drake
University of Leeds, UK

D. Singh
Indian Institue of Technology, Bombay, India

JOHN WILEY & SONS
Chichester · New York · Brisbane · Toronto · Singapore

Copyright © 1996 by John Wiley & Sons Ltd,
Baffins Lane, Chichester,
West Sussex PO19 1UD, England

National   01243 779777
International (+44) 1243 779777

e-mail (for orders and customer service enquiries): cs-books@wiley.co.uk
Visit our Home Page on http://www.wiley.co.uk
or http://www.wiley.com

All Rights Reserved. No part of this book may be reproduced, stored in a retrieval system, or transmitted, in any form or by any means, electronic, mechanical, photocopying, recording or otherwise, except under the terms of the Copyright, Designs and Patents Act 1988 or under the terms of a licence issued by the Copyright Licensing Agency, 90 Tottenham Court Road, London, UK W1P 9HE, without the permission in writing of the publisher.

Other Wiley Editorial Offices

John Wiley & Sons, Inc., 605 Third Avenue,
New York, NY 10158-0012, USA

Jacaranda Wiley Ltd, 33 Park Road, Milton,
Queensland 4064, Australia

John Wiley & Sons (Canada) Ltd, 22 Worcester Road,
Rexdale, Ontario M9W 1L1, Canada

John Wiley & Sons (Asia) Pte Ltd, 2 Clementi Loop #02-01,
Jin Xing Distripark, Singapore 0512

British Library Cataloguing in Publication Data

A catalogue record for this book is available from the British Library

ISBN 0 471 96494 8; 0 471 96496 4 (pbk.)

Produced from camera-ready copy supplied by the authors using LaTeX
Printed and bound in Great Britain by Biddles Ltd, Guildford and King's Lynn
This book is printed on acid-free paper responsibly manufactured from sustainable forestation, or which at least two trees are planted for each one used for paper production.

To Verna

To Kalpana

Contents

1 Some of the history of the concept of sets 1
1.1 Cantor's contribution 1
1.2 Cantor's concept of a set 3
1.3 Paradoxes 5
1.4 Type structures 7
1.5 Culmination into axiomatics 10

2 First-order logic and its use in set theory 13
2.1 Introduction 13
2.2 The syntax of LST 14
2.3 Proofs and derivations in LST 18
2.4 Semantics of LST 19
2.5 Adding new terms 24

3 The axioms of set theory 27
3.1 The Zermelo–Fraenkel axioms 27
3.2 Arguments for these axioms 31

4 Cardinals 37
4.1 Countable sets 37
4.2 Uncountable sets 41
4.3 The arithmetic of cardinal numbers 46

5 Order relations and ordered sets 51
5.1 Orderings 51
5.2 Some properties of ordered sets 53
5.3 Lattices and Boolean algebras 56
5.4 Well-ordered sets 66

6 Developing mathematics within ZFC 71
6.1 The natural numbers 71

6.2	The Peano axioms for the natural numbers	76
6.3	The rational numbers	85
6.4	The real numbers	89
6.5	Ordinals in ZF: basic properties	93
6.6	Transfinite induction	97
6.7	Cardinals as initial ordinals	105

7 The axiom of choice 109
- 7.1 Simple forms 109
- 7.2 The well-ordering theorem 112
- 7.3 Maximal principles and Zorn's lemma 114
- 7.4 Simple consequences of the axiom of choice 117

8 Constructible sets and forcing 127
- 8.1 Gödel's constructible sets 128
- 8.2 The definition of L 133
- 8.3 Reflection principles 138
- 8.4 Properties of L 140
- 8.5 The axiom of choice in L 145
- 8.6 The generalized continuum hypothesis in L 148
- 8.7 Another presentation 153
- 8.8 Forcing models 154
- 8.9 Forcing in practice: the ZFC axioms hold in $M[G]$ 159
- 8.10 Forcing in practice: some models 164
- 8.11 Proofs of the definability and truth lemmas 171
- 8.12 Models for the independence of the axiom of choice 179
- 8.13 A model with a Dedekind-finite set 184
- 8.14 Boolean-valued models: another presentation 189

9 Miscellaneous further topics 193
- 9.1 Introducing variables for classes 193
- 9.2 System VNB 193
- 9.3 System MK 197
- 9.4 Axioms of extent 199
- 9.5 Other presentations of set theory 202
- 9.6 Remarks on the philosophy of mathematics 207

10 Appendix: Some basic definitions 213
- 10.1 Simplest constructions, and variants 213
- 10.2 Ordered pairs, relations, functions, families and sequences 214

References 219

Index 225

Preface

This book had its first origins when the authors met at the European Summer Meeting of the Association for Symbolic Logic, held at Hull in 1986. The suggestion at that time was for a text to meet the needs of MSc students in India and of final year undergraduates and MSc students in Britain; this has evolved into an attempt to write a text which was not a first introduction to set theory (this seemed to be catered for by many undergraduate texts which introduced elements of set theory), but would take students from that point to a point where they would be able to read the many graduate texts which set out the latest researches (or really the researches of the last thirty years or so). Hence the title, intermediate between the simple uses of set theory which are common in university mathematics, and graduate work. It was also intended to introduce some of the foundational aspects which all mathematicians should know of, but many will not want to consider in any greater depth than is presented here. (Of course, it has been argued that working mathematicians should really *not* know anything about foundations, but will do much better to ignore such niceties until they are too old for the real work; even so, this book should make a good introduction at that stage if not earlier!)

It is a real question how far to go in a text which is intermediate in the sense of the present work. The answer has been provided in part by several graduate texts which have not wanted to take the time to present any *proofs* of the basic facts about forcing and constructible sets, but have quoted these basic facts with references to where the proofs can be found. If this book will serve to bring students to the point of being ready for those texts, it will have served some of its purpose.

At the other end, it is hard to find anything of which one can be quite certain that all students of university level mathematics will have met, and we have provided an appendix giving the constructions which we are assuming will be known. Also the presentation of the mathematics for which the set theory is the foundation is begun at a fairly low level in chapter 4, where we do not assume much in the way of background knowledge of cardinals. But it is not really expected that this will be the first time the reader has heard of cardinals and one-to-one correspondences.

We have begun the book with chapters setting out some of the history and some of the formal intent of set theory as a foundation for mathematics. The work introducing constructible sets needs some background knowledge of first order logic, and this is set out in chapter 2, before introducing the axioms in chapter 3. It is not essential for the work in chapters 4 to 7, to have this background knowledge, although it informs the foundational aspects at every step; one way to absorb the work on first order logic is to wait until the point in chapter 8 where it has to be modelled inside set theory, and then to work simultaneously through the definitions in chapter 2 and chapter 8. But we really feel that it is rightly placed before the axioms are introduced.

The work to be formalized, (the actual *mathematics*), starts in chapters 4 and 5. Then the formalization is begun in chapter 6, the axiom of choice is presented in chapter 7, and the more advanced work introducing constructible sets and forcing is in chapter 8. There are a variety of further topics in chapter 9.

Exercises on the work are presented throughout; many of these are intended to make the work useful in teaching. In the nature of the subject, many of these exercises are just the completion of work started in the main text. But others introduce further ideas and results, which are relegated to exercises only to keep the work reasonably concise. We have tried to give references to the literature for these extensions.

We must record our thanks to the publishers, and to their referees for some helpful suggestions; we are sure these will have improved this book. We are also grateful to various students who have read and commented on early drafts, in particular to Sarah Katau.

Frank R. Drake
Dasharath Singh
February 1996

1

Some of the history of the concept of sets

1.1 Cantor's contribution

Georg Ferdinand Ludwig Philip Cantor (1845–1918) is regarded as the father of set theory. He was led to a comparative study of the size (which he called *power*) of infinite sets, by problems relating to trigonometric series; in particular he discovered the importance of the distinction between countable and uncountable sets of real numbers. He introduced the notion of the power of a set (finite or infinite) as a measure of size, and also the notion of transfinite numbers (ordinals and cardinals). This is the remarkable achievement of Cantor's work, published in a series of papers between 1873 and 1897.

Cantor boldly insisted on the existence of infinite sets as mathematical objects, and he regarded infinite sets and transfinite numbers as being on a par with finite sets and ordinary counting numbers. This was not readily accepted by his contemporaries. Cantor seems to have recognized the revolutionary character of this step, and probably envisaged the reaction; it did not deter him.

Interestingly, Cantor showed little interest in justifying the theory of natural numbers using set theory; this is in contrast to other leading contemporaries like Gottlob Frege (1848–1925) and Richard Dedekind (1831–1916), who were very much engaged in justifying set theory as a unifying branch of mathematics. It is worthy of note that they also had considerable success in this programme, notably Dedekind [Ded88]. Eventually it was more or less accepted that the whole of mathematics could be reduced to set theory, and the theory of sets was accepted as an indispensable mathematical tool. But this took a long time; the transition period was to include two notable events, the introduction of the paradoxes, and Zermelo's proof of the well-ordering theorem from the axiom of choice.

In 1897, Burali-Forti published results critical of Cantor's work on ordinals, including what is now known as his paradox. In 1902, Russell gave his much simpler paradox (in a letter to Frege). Many regarded this as a heavy blow to Cantor's theories; but Cantor himself was not much disturbed. From the start Cantor had insisted on the distinction between the *Absolute Infinite* (which, in consonance with the theology of his day, he identified with God, and agreed was beyond mathematical treatment; to form the collection of all things, or all ordinals, would be an attempt to treat the absolute infinite), and the *transfinite* (which was the name he introduced for the size of such sets as the natural numbers or the real numbers, which he insisted were amenable to mathematical treatment, on a par with finite sets). He had thought on the same lines as the Burali-Forti paradox a couple of years earlier, and had communicated with David Hilbert in 1896; later in letters to Dedekind he indicated a modification to his basic definitions which sufficed to keep his theory intact. He transformed the distinction between the transfinite and the absolute infinite into the distinction between consistent and inconsistent "manifolds". (Russell later called these conceivable and inconceivable entities, and this distinction has since been codified, following von Neumann, as the distinction between sets and *proper classes*.) The totality of all ordinals (which is needed for Burali-Forti's paradox), comes under the head of inconsistent totalities (nowadays we say it is a proper class), and Cantor criticized Burali-Forti for overlooking this distinction, and also noted that Burali-Forti had misstated the definition of a well-ordered set. (See [YY29] for a discussion of Burali-Forti's mistakes and of the paradox.)

A related problem arises with Frege, who wanted to take all classes as objects which could be members of sets; we shall consider this later.

Certainly if the distinction is maintained, and the inconsistent manifolds rejected, then the paradoxes such as those of Burali-Forti and Russell cannot arise. But Cantor could only give some indications about the distinction; for example: "Two equivalent manifolds are either both sets or both inconsistent"; "every subset of a set is a set" ([Can32] p. 444). He could not carry out the distinction in full detail, and the problem does reflect the inadequacy of our approaches to this day. (It connects with the problem of large cardinal axioms, mentioned in 9.4).

Cantor did not work with a codified system, and was not interested in doing so, although almost all the axioms of set theory can be derived from a close analysis of his proofs. But in 1904, Ernst Zermelo published his proof that every set can be well-ordered (a theorem which Cantor had first assumed to be true, and later tried to prove from simpler assumptions, but finally left as an open problem). Zermelo gave his proof first, in [Zer04], in a form which was not completely formalized, and many mathematicians were confused about several of the ideas. It was this proof which most clearly brought to attention the possible variations in the meaning of "existence" in mathematics. These range from an outright construction, with a finite set of instructions, on the one hand, through to a proof which gives no possibility of constructing the object in question, but merely shows that it has to exist (perhaps by giving a contradiction if it didn't exist), on the other. The well-ordering of the real numbers, which Zermelo's theorem implied, could not in any sense be constructed; and this was not readily accepted by many mathematicians. This led to Zermelo's axiomatization; in [Zer08] he published a second version of his proof, this time noting carefully the axioms or fundamental properties of sets which he was using in his proof, including of course the axiom of choice. (For a much fuller account of this history, the reader is recommended to [Moo82]; for a fuller account of Cantor's work, to [Hal84]. Many of the original letters and papers are translated in [vH67].)

In the next few sections we consider some of these points in a little more detail.

1.2 Cantor's concept of a set

Cantor defines *set* in the following words:

"Unter eine 'Menge' verstehen wir jede Zusamenfassung M von bestimmten wohlunterschiedenen Objekten m unserer Anschauung oder unserer Denkens (welche die 'Elemente' von M genannt werden) zu einem Ganzen." ([Can32] p.282)

Translated: "By a 'set', we are to understand any collection M of definite and distinct objects m of our intuition or thought (which will be called the 'elements' of M) into a whole."

The requirement that the objects be <u>distinct</u> probably would be intended to mean that for a given pair of objects to appear as elements in a set, one must be able to determine whether they are different or the same. In fact there is some confusion possible, about what to do about repetitions of the same element. Eventually we shall adopt, following Zermelo, the <u>axiom of extensionality</u> which effectively means that repetitions will be ignored, and a set such as $\{a, a, b, c, c, c\}$ will be regarded as exactly the same as $\{a, b, c\}$. Such a theory is then said to deal with <u>extensional</u> sets. In fact there are many occasions in practice when one does want to consider repetitions, and so (apparently) reject this identification. Thus in order to talk about rational numbers, one sometimes wants to regard $\frac{3}{9}$ and $\frac{1}{3}$ as distinct, and sometimes as identical. In making observations, one may record the same reading several times, and for statistical purposes it will be important to note the frequency of occurrence. But this sort of example also reminds us of the normal way of dealing with repetitions—one supplies a label (such as the observation number) so that different occurrences are given different labels. The labels will most often be natural numbers (observation one, observation two, ...) and this means that we are thinking of a *sequence* rather than a *set*. Since there is a quite adequate way of dealing with sequences or with labels within the theory of extensional sets (see 10.2.4), we can quite reasonably restrict our attention to the case where repetitions are ignored.

(Note that sets with repetitions have been called _multisets_ and treated in the literature. But we shall not deal with them here.)

The requirement that the objects be <u>definite</u> is usually taken to imply that one should be able (at least in principle) to tell whether a particular object is in the set or not; thus we are not dealing with _fuzzy_ sets (where we may be unsure whether an object is in or not). Again this is really a decision to restrict our attention to the simplest case, and it is usually expressed by working within classical logic, where we take the proposition $m \in M$ (which expresses: object m _is an element of_ set M, or more simply m _is a member of_ M, or m _belongs to_ M) as being either true or false, with no other value allowed. For example, let the set of all rational numbers be taken as the set of representations of the form $\frac{p}{q}$, where p and q are integers with highest common factor $(p,q) = 1$ and q is positive. This set is well-defined because, given any number, one could in principle decide whether or not it is in the set (even though in a particular case such as 2^e or $\sqrt{2}^{\sqrt{2}}$ we ourselves may not know which, nor any way to decide which—that is not taken to matter).

{L.E.M.}

However it is worth noting that the demand for definiteness cannot always mean that the sets are given by some specific formal definition. For example we shall want to consider sets of the type $C = \{a_1, a_2, a_3, \ldots\}$ where each $a_n \in A_n$ and is an arbitrary choice, from the family of sets $\{A_1, A_2, A_3, \ldots\}$, each having more than one element. This is precisely what is done when one accepts the <u>axiom of choice</u>, and one outcome of modern independence proofs (as mentioned in 8.12) shows just how and when such families can arise for which no set of the type C (which is called a _choice set_ for the family) can be defined.

The status of the axiom of choice is set out in chapter 8, which gives the results of Kurt Gödel ([Göd38]) and Paul Cohen ([Coh63a]) on the subject. The discussion of the axiom itself is in chapter 7.

As regards the nature of the objects themselves, "objects of our intuition or thought": Cantor himself treated such objects as the points, lines and planes of Hilbert's geometry, and used them as building blocks for the construction of further objects of thought. In later developments, Cantor's objects of intuition were treated as _individuals_ (in German _Urelemente_), and objects of thought as _sets_; and it is worth noting that the strong form of the axiom of extensionality which we shall adopt (following Zermelo), is not really in Cantor's spirit: it takes the objects of thought (sets) alone as sufficient, and leaves out individuals. This "sufficiency" is only for the development of mathematics within set theory, however, and in many other applications individuals _are_ wanted (and the axiom of extensionality is weakened accordingly; see 3.2.3(1)). Other ways of treating individuals have also been suggested, e.g. by Quine

HISTORY OF SETS

[Qui37] (see 9.6.5), which weaken other axioms.

Finally we examine Cantor's commitment to a set being a "collection into a whole". It seems that this may have originally been taken by others as the following two parts: first, every set has a defining property (predicate), and second, corresponding to every property (predicate) there is a set consisting of all the objects satisfying that property. Both have now to be viewed false (see Thoralf Skolem, in [Sko23], [Sko30] and other writings). We commented on the first point above in connection with definiteness, and in fact Cantor must not have taken set construction in terms of predicates alone, simply because, with only a countable number of predicates of a natural language in hand, (Cantor was working in a natural language), an uncountable set of sets would not have been conceived of.

But the second point gave more trouble. We have seen above that Cantor insisted on the distinction between transfinite sets, which can be considered as collected into a whole, and other collections which corresponded to the absolute infinite and could not be so considered. The neglect of this distinction was widespread; Frege was probably the most notable to attempt to consider all classes on a par. This is now called the *general principle of comprehension*, and can be stated as: given any property $P(x)$, there is a set whose elements are exactly those entities having that property. In symbolic form, we introduce the *abstraction term* $\{x \mid P(x)\}$, with the defining formula:

$$\forall x[x \in \{x \mid P(x)\} \Leftrightarrow P(x)]. \tag{$*$}$$

In a letter to Frege of 1902, see [vH67], Bertrand Russell (1872–1970) gave his paradox concerning this principle:

1.3 Paradoxes

1.3.1 Russell's paradox

Let $P(x)$ stand for $x \notin x$ and let $y = \{x \mid P(x)\}$. Now, if $y \in y$ then by $(*)$, $P(y)$ must hold, i.e. $y \notin y$ must hold. But also, if $y \notin y$ holds then by definition of y and $(*)$, $y \in y$. In words: the general principle of comprehension implies that the collection of all sets which are not members of themselves is a set. And this set has the property that it is an element of itself if and only if it is not an element of itself. So we have a contradiction.

Two related paradoxes were given later:

1.3.2 Paradox of groundedness

(See [Mon55].) Let a set x be called <u>groundless</u> if and only if there exists a sequence of sets $x_1, x_2, \ldots, x_n, \ldots$ (not necessarily all distinct) such that

$$\ldots \in x_{n+1} \in x_n \in x_{n-1} \in \ldots \in x_2 \in x_1 \in x.$$

(If we think of the members of x as in some sense lower than x, then we can descend forever, starting from a groundless set x.) A set is called <u>grounded</u> if it is not groundless. Let $P(x)$ now stand for "x is grounded" and again let $y = \{x \mid P(x)\}$. Now if y is grounded, then $y \in y$ by $(*)$ and hence

$$\ldots \in y \in y \in y \in \ldots \in y \in y \in y,$$

and this implies that y is groundless, hence a contradiction. But again, if y were groundless, then by definition there would exist a sequence $y_1, y_2, \ldots, y_n, \ldots$ such that

$$\ldots \in y_{n+1} \in y_n \in y_{n-1} \in \ldots \in y_2 \in y_1 \in y.$$

This implies that y_1 is also groundless. But then $y_1 \in y$ contradicts the definition of y. (A more general definition of groundedness is in exercise 1.5.1(1).)

1.3.3 Paradox of non-circular sets

Let a set x be called <u>n-circular</u> (where n is a natural number) if and only if there exist sets $x_1, x_2, \ldots, x_{n-1}$ (not necessarily all distinct) such that

$$x \in x_{n-1} \in x_{n-2} \in \ldots \in x_2 \in x_1 \in x.$$

Also let a set be called *circular* if and only if it is n-circular for some $n \geq 1$, and *non-circular* if it is not circular. Now let $P(x)$ stand for "x is non-circular", and let $y = \{x \mid P(x)\}$.

If y were non-circular, then by definition of y and $(*)$, $y \in y$ so that y is 1-circular, hence circular, a contradiction.

But if y were circular, by definition of circularity there exist sets $y_1, y_2, \ldots, y_{n-1}$ such that

$$y \in y_{n-1} \in y_{n-2} \in \ldots \in y_2 \in y_1 \in y.$$

For $n = 1$ we have $y \in y$, so by definition of y, y should be non-circular, a contradiction.

For $n > 1$, we have from above

$$y_1 \in y \in y_{n-1} \in \ldots \in y_2 \in y_1,$$

which implies that y_1 is circular. But also $y_1 \in y$, so by definition of y, y_1 should be non-circular, again contradiction.

In a sense, the earliest paradox was probably:

1.3.4 Cantor's paradox

Let $P(x)$ stand for "x is a set" and let $y = \{x \mid P(x)\}$, i.e. y is the set of all sets. Then every subset of y is also an element of y (since it is a set). So the power-set of y (the set of all subsets of y) is a subset of y. This contradicts Cantor's theorem (see 4.2.4), which says that the power-set of any set is strictly larger than the set itself (and so cannot be a subset of it).

Note that Burali-Forti's paradox will be presented in 6.5.7, and some other results we present could be regarded as paradoxes of just the same sort. They all illustrate the fact that the general principle of comprehension is simply too strong. In each case it is possible to view the collection of objects satisfying $P(x)$ as being very large, too large to form a set. Cantor had almost ceased from active work by the time Russell's paradox, the simplest, became known; but his original intention, that such properties $P(x)$ which lead to contradiction when used in the comprehension principle, should be considered as corresponding to the absolute infinite and not amenable to mathematical treatment, has not been bettered. However, this does not give us a direct way to decide what *can* be regarded as a set, or object, and we shall consider some of the solutions to this problem that were given.

1.4   Type structures

Russell's first reaction to the paradox 1.3.1, which he discovered on reading Frege's work, was the "vicious circle principle" which declared all predicates of the form $x \in x$, $x \in y \in x$, etc. (as in 1.3.3) to be meaningless. But then in trying to reinstate the essential features of the richness of Cantor's set theory, he found that he had still admitted circular sets: he defined 1 as the set of all singletons, 2 as the set of all doubletons, and so on (where a singleton is intuitively understood as a set containing one element, a doubleton as a set containing two elements, and so on). But then he defined singleton $\{1\}$ as the set x such that $\forall y[y \in x \Leftrightarrow y = 1]$ and then $\{1\} \in 1$ by definition, so that $1 \in \{1\} \in 1$, a contradiction to the vicious circle principle.

The next step he took was in conjunction with A. N. Whitehead, in *Principia Mathematica* [WR10]. Essentially they there introduced the notion of <u>types</u> or <u>levels</u>: if x and y are two collections and $x \in y$, then x and y must be of different types; y must be at one *level* higher than x. At the same time, they divided the sets of the same level into <u>orders</u>; this produced what became known as the <u>ramified theory of types</u>, and this was developed in *Principia Mathematica* into a basis for mathematics. But it has generally come to be regarded as too complex (particularly because of the extra ramification into orders, which were introduced to avoid the so-called semantic paradoxes such as those of Berry and Richard (see [vH67]), and also not at all intuitive (because of the extra axioms needed to give the strength required for much of mathematics; in particular the *axiom of reducibility* never received intuitive justification). But the notion of levels (or *types*) has proved important, and we describe Russell's notion of simple types here.

1.4.1 The simple types

The sets of level 0 are called individuals (they correspond very well to Cantor's "objects of intuition"; they have no members, but there are assumed to be many of them);

the sets of level 1 are collections whose members are all individuals (i.e. of level 0);

the sets of level 2 are collections whose members are all of level 1;

the sets of level 3 are collections whose members are all of level 2;

$$\vdots$$

the sets of level $n+1$ are collections whose members are all of level n;
and so on.

Now a set "1" can be formed consisting of all the singletons in level 1; but this set "1" itself will be at level 2, and $\{1\}$ will be at level 3, and we have certainly avoided the violation of the vicious circle principle noted before. The intuitive idea described by these levels is nowadays known as the *simple type structure* (*simple* because we have omitted any reference to the *orders* which Russell introduced, and which give the *ramified* type structure). This simple type structure is an important concept, much studied in modern logic; it forms an appropriate setting for many studies. However it is not sufficiently rich by itself to form a foundation for mathematics, and it is easy to see some of its deficiencies from the point of view of Cantor's definition of a set. In particular there is no clear reason to rule out sets with members from different levels. If a and b are individuals at level 0, then $\{a\}$ and $\{b\}$ are at level 1, and $\{\{a\},\{b\}\}$ is at level 2, but nowhere does $\{a,\{a\}\}$ or $\{a,\{b\}\}$ occur (sets with two members, one a, the other $\{a\}$ or $\{b\}$). Yet it seems no more problematic than $\{\{a\},\{b\}\}$. Hence various mathematicians (Mirimanoff may have been the first; [Mir17]) have suggested what is now known as the *cumulative* type structure.

1.4.2 The cumulative types

The sets of level 0 are individuals (just as before);
and
the sets of level 1 are collections of individuals (just as before);
but now

HISTORY OF SETS 9

the sets of level 2 are collections, all of whose members are of level 1 *or* level 0;

the sets of level 3 are collections, all of whose members are of levels 2, 1 or 0;

and in general

the sets of level n are collections, all of whose members are of levels $< n$; and so on.

One of the complications of using the simple type structure as a foundation for mathematics was the following: Russell wanted to define the numbers as sets, and for example "1" was to be the set of all singletons—but at which level? In particular the empty set (which is an obvious choice for the number zero) could be considered as existing at each level, and one must be careful to say which level one is thinking of.

But now note that the cumulative type structure makes no problem of this feature: if a set exists at some level, that same set is available at every higher level (to be a member of some new set which exists at that higher level), and it could be considered (if we want to so consider it) as being formed anew at *every* subsequent level. But for each set there will always be a *first* level at which it exists (namely the first level after all of its members exist), and this first level is nowadays called the <u>rank</u> of the set. It turns out to be a very useful concept, and we discuss it further in 6.6.13(3).

In fact all of our subsequent work could be considered as based on the cumulative type structure, and all the axioms we give in chapter 3 can be given intuitive justification in terms of this structure; we could say that they are intended to give a formal description, or *theory*, of this structure. We shall even make the severest restriction possible about urelements: we shall assume there are none (or equivalently we could consider that the empty set is the only urelement; see 6.6.13(3)). In some sense this is a good way to regard modern set theory. But it must be noted that it is not the historical way; it seems clear that Zermelo and Cantor did not conceive of sets in this way. Mirimanoff seems to be the earliest to have ideas along these lines. F.P. Ramsey [Ram26] noted that Russell's theory did not need the ramifications of orders in order to avoid paradoxes; and J. von Neumann [vN25] introduced some elements of the notion (but working with functions rather than sets). Insistence on the importance of the notion of rank seems to come with Dana Scott [Sco57]. (Some of the early work was formal, and *theory* rather than *structure* was studied, which makes comparison with later viewpoints more difficult.)

1.4.3 Resolution of the paradoxes

Note how the paradoxes are explained in the context of this cumulative type structure: a property $P(x)$ only determines a set $\{x \mid P(x)\}$ if there is some level such that every object x satisfying $P(x)$ already exists before that level (and no others exist at that or later levels). This is clearly not so for the properties $x \notin x$ and $x = x$ which will apply to every set at every level; and in fact all of the paradoxes 1.3.1–4 are describing the collection of all sets. From the point of view of the cumulative type structure, that does not exist at any level. (Note that even if there is a *last* level, the collection of all sets will not exist, but only the collection of all sets existing *before* the last level—which will itself be a new set at this last level.)

Collections which have members from arbitrarily high levels are now called <u>*proper classes*</u>, and we resolve any of the known set-theoretic paradoxes if, when we form collections $\{x \mid P(x)\}$, we regard them as classes, not eligible for membership of anything, unless or until we have a proof that they are <u>*bounded in rank*</u>, i.e. that there is a level such that every x with property $P(x)$ occurs before that level.

It is interesting to note that Burali-Forti's paradox, not yet presented (see 6.5.7), does differ from those in 1.3.1–4 in that it describes a collection which, while not containing every set, still is unbounded in rank. There must clearly be many such, but only one of the early paradoxes did not refer to the whole universe of sets (from the point of view of the cumulative type structure).

1.5 Culmination into axiomatics

There have been many outstanding developments in mathematics which are mainly due to paradoxes (even though nobody wishes it so!); the development of set theory is no exception. The codification of set theory which we shall present in chapter 3 has emerged as a powerful tool and a satisfactory basis, even though foundational problems are not completely resolved (nor can we expect them to be). This task was first taken up by Ernst Zermelo ([Zer08], [Zer30]), and then by Mirimanoff ([Mir17], [Mir20]), Abraham Fraenkel ([Fra22a], [Fra22b]), Thoralf Skolem ([Sko23], [Sko30]), Paul Bernays ([Ber54]), Kurt Gödel ([Göd40]), John von Neumann ([vN25], [vN28a]), W. Ackermann ([Ack56]), Azriel Levy ([Lev59]), Willard van Orman Quine ([Qui63]), and others.

Not all of these workers have formulated their systems of axiomatics using exactly the same interpretation of Cantor's concept of set; but we shall only make very brief mention of other interpretations, in 9.5.

1.5.1 Exercises

(1) Grounded sets. The definition of "grounded" in 1.3.2 is the simplest way to think of the matter, but it depends on the axiom of choice if we are to assume that descending sequences will exist whenever they *could* exist. [This is really an application of dependent choices as in 7.4.3.] A definition

of *grounded* which does not assume the axiom of choice is:

A set x is *grounded* if and only if $\forall a[x \in a \Rightarrow \exists y \in a(y \cap a = \emptyset)]$.

Show that this implies the definition used in 1.3.2, and that this definition also gives the paradox.

[Intuitively the definitions are equivalent. If $x \in a$ and $\forall y \in a(y \cap a \neq \emptyset)$, then $x \cap a \neq \emptyset$ so let $x_1 \in x \cap a$; then $x_1 \cap a \neq \emptyset$ and let $x_2 \in x_1 \cap a$; then $x_2 \cap a \neq \emptyset$ and let $x_3 \in x_2 \cap a$, and so on. This gives our descending sequence, but by an intuitive use of dependent choices. For the other direction, think of a as the set of all members of the descending sequence starting from x. Note that a use of the paradox in this form, or really of the argument which gives the paradox, is in 9.5.]

(2) *k*-circular sets. The paradox in 1.3.3 used the predicate "non-circular"; show that a paradox can arise from the predicate "non-*k*-circular" for any fixed $k \geq 1$.

[Russell's paradox would then be the case $k = 1$.]

cummulative type structure; proper classes are not allowed as members of sets

2

First-order logic and its use in set theory

2.1 Introduction

Logic developed over a much longer history than set theory, but in some ways the two have been intertwined throughout that history. The outcome might be said to be two separate notions: first-order logic, and set theory. But we cannot present modern set theory without making use of first-order logic. It will be an important part of our work to check what axioms are needed for the deductions we make, and to show that certain results really do follow from our axioms. For this we need to use first-order logic. In effect we are making use of the results of the study of logic, in particular the completeness theorem for first-order logic, in order to present set theory in a completely unambiguous way.

This is partly a matter of communication: by presenting the results in a completely formal way (i.e. within a first-order language), we can be quite certain that we have in fact proved what we claim to have proved, and that we can communicate that fact to others. We know that we could even program a computer to check that fact—and indeed this has been done, and much of mathematics, including some of set theory, has been so checked. (But this was not done because there was serious doubt about the mathematics; it was done rather as an exercise in computing, and so that the methods could then be applied in cases where there *was* doubt, such as in checking the correctness of computer programs.)

In fact we can think of set theory today in the following way: on one side, a heuristic notion, the cumulative type structure, is intuitively described. On another side, a formal, first-order language, LST (for: Language of Set Theory), is set up, with one primitive notion, the membership relation $\in$. Then various sentences of this language are singled out as axioms, and these are compared with the intuitive structure to see whether they hold or not; in this way our intuitions are sharpened, and the axioms are sorted into some which we definitely accept and others which are tentative. Thus the Zermelo–Fraenkel axioms and the axiom of choice will be furnished with arguments as to why we should accept them; while large cardinal axioms will have only more tentative reasons, and axioms such as the axiom of determinacy or the axiom of constructibility may be regarded as true only for a suitably changed heuristic structure (usually some restricted part of the whole structure).

In this work, the one clear, unambiguous part is the part within first-order logic: from certain axioms we can prove certain conclusions. For example, we shall set up definitions for numbers (natural numbers, real numbers, ...) and show that, assuming the axioms of set theory, these defined notions have the properties we expect. (This is the clearest sense in which set theory forms a foundation for mathematics, and is carried out in chapter 6.) This is the sort of work about which there is no doubt, and which could be checked by a computer: within first-order logic, from *these* axioms follow *those* conclusions.

Note that in this sense there is much that can be said, quite unequivocally, about such a doubtful notion as the axiom of determinacy (which contradicts the axiom of choice, and so seems very unlikely to be true about the cumulative type structure as a whole); we can say without any doubt, for example, that *if* the axiom of determinacy holds, *then* all sets of reals are Lebesgue measurable.

But there are many things we *cannot* do with the same freedom from doubt, and the clearest such is to prove that our axioms are consistent, i.e. free from contradictions—or more specifically, to show even one statement that cannot be proved from our axioms. (It is a simple result of classical first-order logic that any system of axioms is inconsistent if and only if *every* sentence can be proved from them.) Gödel's incompleteness theorem shows that, for any theory strong enough (such as even the simplest set theory), there cannot be a rigorous proof of consistency without first assuming the consistency of a stronger system. So the *only* way open to us, to convince ourselves or others that the axioms we are using, are free from contradictions, is to give heuristic arguments, such as by exhibiting the heuristic notion of the cumulative type structure and giving reasons why our axioms should be true for that notion. This is what we shall be attempting, for the Zermelo–Fraenkel axioms, when we present them in chapter 3 and explain their meaning.

In fact it turns out that we can give other arguments if we change the language by adding other primitives, in particular ones that directly concern the notion of level or rank, and some of this will be presented in 9.5. This might be considered a better justification, at least for some of the axioms.

2.2 The syntax of LST

We present first a very simple language, almost the very minimum which could be used, then later we extend the language by definitions so that it becomes

FIRST-ORDER LOGIC AND ITS USE IN SET THEORY

rich enough to present, in effect, most of the work of modern mathematics. This means that a large part of the later work will consist of definitions, together with lemmas to prove these have the properties that we already know and are familiar with. But we start with the very simple language because we can then examine more easily the full statement of our axioms, and compare them with the heuristic picture; the later definitions will involve no further assumptions, but must be justified from the axioms given.

The basic components of a first-order language are: first, the _terms_ (which represent the objects, or things we are talking about). Second, the _predicate symbols_ (which represent the simplest statements we can make, about the objects; they are used to form atomic formulas). Third, the _logical connectives_ and the _quantifiers_ (which are used to make more complicated statements from simpler ones). From these basic components we build _formulas_.

2.2.1 Terms (or object symbols)

These will be variables; formally we shall use $v_1, v_2, v_3, \ldots$ as our variable symbols. We shall then write $a, b, c, \ldots, x, y, z, \ldots, t, u, v, \ldots$ and other letters, with and without suffixes, as what are called _meta-variables_ to stand for any variable; this is a convenience, e.g. in presenting definitions such as 2.2.3 below.

For our simplest language, variables will be the only terms. Later we shall want to introduce many _defined_ terms, in particular _constant symbols_ such as $\emptyset$ for the empty set, and _abstraction terms_ of the form $\{x \mid P(x)\}$ —we know that this must be done with care, in view of the paradoxes, and we shall give in 2.5 some sufficient conditions for adding defined terms.

2.2.2 Predicate symbols (or relation symbols)

We use only two binary predicate symbols, $\in$ and $=$.

The symbol $=$ will be used to denote the identity predicate, which is often regarded as a _logical predicate_; all other predicate symbols, in our case $\in$, which we use to denote the membership relation, are called _non-logical_. The distinction shows in 2.4.1, and is not important until then. We later introduce many further predicate symbols by definitions; in this case there are no complications, since introduced predicate symbols can always be regarded as simply abbreviations.

2.2.3 Atomic formulas

An atomic formula is a string of symbols of the form $x = y$ or $x \in y$ where x and y are any terms (so for the moment, just variables).

These will express the basic propositions of our theory, the idea being that when suitably interpreted, an atomic formula will be either true or false. But

---

Atomic formula are wffs containing a single predicate & no connectives

that is part of the semantics in 2.4; the formal syntax knows nothing of that, just as a computer knows nothing of the meaning of symbols.

2.2.4 Logical connectives

The symbols $\neg$ and $\Rightarrow$ will be taken as our only logical connectives.

They are intended to denote negation (or "not") and material implication (or "if ... then") respectively; all other of the classical connectives, such as $\wedge$ (conjunction, or "and"), $\vee$ (disjunction, or "or"), $\Leftrightarrow$ or $\equiv$ (bi-implication or equivalence, or "if and only if"), will be taken as abbreviations for their appropriate definitions in terms of $\neg$ and $\Rightarrow$. (Note that we use $\Rightarrow$ in preference to $\rightarrow$ or $\supset$ for material implication, since $\rightarrow$ already has two mathematical senses, namely in mappings 10.2.3 or as "tends to a limit" when we present the real numbers in chapter 6; and $\supset$ suggests "contains as a subset", and is sometimes used for that.)

2.2.5 Quantifier symbol

$\exists$ (for "there exists", the existential quantifier).

We shall take the universal quantifier $\forall$, "for all", as being defined from $\exists$, and also such quantifiers as $\exists^1$, "for just one".

2.2.6 Formulas

Formulas (or *well-formed-formulas*) are now given by an inductive definition in the meta-language, and they use also the punctuation symbols, right and left brackets [and].

  (i) Atomic formulas are formulas.
  (ii) If φ and ψ are formulas, then $\neg\varphi$ and $[\varphi \Rightarrow \psi]$ are formulas.
  (iii) If φ is a formula and x is a variable, then $\exists x \varphi$ is a formula.

We are using here meta-variables φ, ψ, χ, possibly with primes or suffixes, to denote arbitrary formulas.

2.2.7 Free and bound variables

We shall need the following definitions about formulas:

  (i) In the formula $\exists x \varphi$, the *scope* of the quantifier $\exists x$ is φ. The scope does not change when formulas are combined.
  (ii) For any occurrence of a variable x in a formula ψ, that occurrence is *free* unless it is within the scope of a quantifier $\exists x$, in which case it is *bound*.
  (iii) A formula with no free variables is said to be a *closed* formula or a *sentence*.

FIRST-ORDER LOGIC AND ITS USE IN SET THEORY

The distinction between free and bound variables is important; put simply, the free variables in a formula will denote objects that the formula is *about*, in the sense that it will hold for some and not for others (in general); they are sometimes called *parameters*, and changing them will change the meaning of the formula (it would then be saying something about different objects). But a bound variable is acting very differently; it is perhaps best thought of as a place marker. We might verbalize the formula $\exists x \varphi(x)$ as saying: there is some *thing* such that φ is true of that *thing*; and whichever variable we use to denote the possible *thing*, it will not alter the meaning of the formula, which will remain true or false. At least, this will be so if we avoid changing this bound variable to another which occurs free in φ already; the following example shows what could then happen: $\exists x\, x \in y$ "says" that y has some member (when we interpret $\in$ as membership and assume that y is interpreted as an object—we might then think of y as the name of an object, called y). $\exists z\, z \in y$ says just the same. But $\exists x\, x \in z$ says something different—namely that the object called z has some member. And most important: $\exists y\, y \in y$ is now a closed formula; it says nothing about y or any other particular object. Now it says: "there is some object which is a member of itself" (which we shall note as false about the cumulative type structure, though it could be true of some other structure).

Because of this last distinction, we shall usually adopt the convention of writing formulas using meta-variables such as $a, b, c, \ldots$ for free variables (parameters), and $x, y, z, \ldots$ for bound variables, so that we shall not so easily make the mistake of confusing free and bound variables. But we do not make this a formal requirement; we could, but we are aiming to present what is (formally) a very simple system, with as few complications as possible.

2.2.8 Some defined symbols

The other connectives and quantifiers are treated as abbreviations. These are the definitions we shall use:

(i) $[\varphi \vee \psi] := [\neg \varphi \Rightarrow \psi]$

(Note that we use $:=$ as the symbol for definitions, at the meta-level where we are describing the formal language.)

(ii) $[\varphi \wedge \psi] := \neg[\varphi \Rightarrow \neg\psi]$
(iii) $[\varphi \Leftrightarrow \psi]$ or $[\varphi \equiv \psi] := [[\varphi \Rightarrow \psi] \wedge [\psi \Rightarrow \varphi]]$
(iv) $\forall x \varphi := \neg \exists x \neg \varphi$
(v) $\exists^1 x \varphi(x) := \exists y \forall x [\varphi(x) \Leftrightarrow x = y]$ where y is not free in $\varphi(x)$.

We shall see in 2.4.2 that these definitions will make these defined connectives have the right meaning ("or", "and", "if and only if", "for all", and "for just one", as promised in 2.2.4 and 2.2.5).

2.2.9 Omitting brackets

We shall rarely write the formula itself, more often an indication of how the formula should be reconstructed. So we shall often omit brackets in accordance with the convention that $\Rightarrow$ and $\Leftrightarrow$ bind less closely than $\vee$ and $\wedge$. Then $\varphi \vee \psi \Rightarrow \chi \wedge \theta$ abbreviates $[[\varphi \vee \psi] \Rightarrow [\chi \wedge \theta]]$ and nothing else. Also we assume that $\Rightarrow$ associates to the right, so that $\varphi \Rightarrow \psi \Rightarrow \chi$ means $[\varphi \Rightarrow [\psi \Rightarrow \chi]]$. We shall also vary the size of brackets to improve readability, and use parentheses $(,)$ as well; we can formally regard these as all representing the same symbol. We usually omit the outermost brackets.

2.2.10 Universal closure

If $x_1, \ldots, x_n$ are all the free variables of φ, we call $\forall x_1 \ldots \forall x_n \varphi$ the _universal closure_ of φ.

If a formula has free variables, we shall see that its universal closure is a very natural sentence to associate with it.

2.3 Proofs and derivations in LST

We are not trying to teach first-order logic for its own sake here, but simply presenting the facts about first-order logic which are the reasons why it is used in presenting set theory. These facts use the notions of logical axioms and rules of inference, and we give some possible definitions here for definiteness. But we shall not use these, we shall only quote the basic facts about them.

2.3.1 Logical axioms

A formula is a logical axiom if it has the form

(1) $\varphi \Rightarrow [\psi \Rightarrow \varphi]$
(2) $[\varphi \Rightarrow [\psi \Rightarrow \chi]] \Rightarrow [[\varphi \Rightarrow \psi] \Rightarrow [\varphi \Rightarrow \chi]]$
(3) $[\neg \varphi \Rightarrow \neg \psi] \Rightarrow [\psi \Rightarrow \varphi]$
(4) $[\forall x \varphi] \Rightarrow \varphi$ if x is not free in φ
(5) $[\forall x \varphi(x)] \Rightarrow \varphi(y)$ where y is any variable _free for x in_ $\varphi(x)$, i.e. such that x does not occur free in $\varphi(x)$ in the scope of a quantifier $\exists y$ or $\forall y$
(6) $\forall x[\varphi \Rightarrow \psi] \Rightarrow [\varphi \Rightarrow \forall x \psi]$ if x is not free in φ
(7) $x = x$
(8) $x = y \Rightarrow [\varphi(x) \Leftrightarrow \varphi(y)]$ for any formula φ, where $\varphi(y)$ has y at _some_ places (not necessarily all) where $\varphi(x)$ has x free, and not within the scope of $\exists y$ or $\forall y$.

FIRST-ORDER LOGIC AND ITS USE IN SET THEORY

2.3.2 Rules of inference

(1) <u>Modus ponens</u>: from φ and $\varphi \Rightarrow \psi$ infer ψ
(2) <u>Generalization</u>: from φ infer $\forall x \varphi$

2.3.3 Deductions and proofs

Using these we can now define the (syntactic) notions of *deduction* and *proof*.

A sequence of formulas $(\varphi_i)_{i \leq k}$ is a <u>*deduction* from</u> Γ (where Γ is a set of formulas, called <u>hypotheses</u>) if, for each $i \leq k$, } φ is then a consequence of Γ

(i) φ_i is an axiom; or
(ii) φ_i is a member of Γ; or
(iii) φ_i can be inferred from earlier members of the sequence by modus ponens or generalization.

We shall write $\Gamma \vdash \varphi$ (Γ *yields* φ, or φ can be <u>deduced</u> from Γ), if there is a deduction $(\varphi_i)_{i \leq k}$ from Γ where φ is the last formula φ_k; and if Γ is empty we call $(\varphi_i)_{i \leq k}$ a *proof* and write $\vdash \varphi$. (For examples of such completely formal derivations and proofs see the references on first-order logic, e.g. [Ham88], [Men64], [End72], [Sho67].)

The importance of these rules and axioms lies in the fact that they are } p.22
sound and *complete*; and these notions involve the semantics of LST.

2.4 Semantics of LST

As we have hinted already in several places, the formal language is not set up for its own sake alone, but is intended to have a meaning, which is summed up in the *semantics*. We set this out in some detail; when we come to present the constructible sets in chapter 8, we shall have to return to this definition and model it within our set theory (just as we shall be modelling the natural numbers within our set theory in chapter 6).

The semantics is given in terms of an *interpretation*, which has two parts, a *structure* and an *assignment* in that structure. A *structure* can be taken as something similar to the cumulative type structure which we introduced in chapter 1; but at this point we want to admit any structure of the right shape. For interpreting LST, we need just two parts: a *domain* D and a binary relation E on that domain. The domain (or *universe*) is just a non-empty collection of objects, which will be the range of our variables; and an *assignment* in the structure will simply be a function v from the variables into D. E will be some collection of ordered pairs of objects from D, and we say two objects σ, τ from D are in the relation E if and only if the ordered pair $\langle \sigma, \tau \rangle$ is in the collection E. E will interpret $\in$ (our <u>membership</u> relation).

The relations E model logical predicates on D
assignment v maps free vars. in wff φ to objects in D

If we are thinking of the cumulative type structure, D will be the collection of all sets (existing at any stage), and E will be the collection of all ordered pairs of sets $\langle \sigma, \tau \rangle$ for which σ is a member of τ. The effect of our definition of satisfaction (or truth) will be that $x \in y$ will be interpreted as true just when the assignments $v(x) = \sigma$ and $v(y) = \tau$ are such that $\langle \sigma, \tau \rangle$ is in E; for the cumulative type structure, this is when σ is a member of τ. This may seem a long-winded way to say it, but it should be clear when we have given the definitions that this is exactly what we mean when we say that the formal symbol $\in$ represents the membership relation.

As noted, we want to set out meaning for our language when we are given an interpretation, i.e. a structure (D, E) and an assignment v. We do this by defining the satisfaction relation

$$(D, E) \models \varphi\,(v)$$

(read: structure (D, E) *satisfies* formula φ of LST at the assignment v; or, formula φ is *true* in structure (D, E) at assignment v). This is defined by induction on the construction of φ as follows:

Definition 2.4.1 (i) If φ is an atomic formula $x = y$, $(D, E) \models x = y\,(v)$ if and only if $v(x) = v(y)$ (i.e. if and only if x and y are assigned to the same object).

(ii) If φ is an atomic formula $x \in y$, $(D, E) \models x \in y\,(v)$ if and only if the pair $\langle v(x), v(y) \rangle$ is in the collection E (or we may say: in the relation E).

Note that this clause gives effect to the intention stated above, that E interprets the membership relation $\in$. It is also at this point that we can see the difference between the logical symbol $=$ (which is interpreted the same way in every structure, by the identity relation) and the non-logical symbol $\in$ (which will have whatever interpretation E happens to give it, and will vary from structure to structure).

(iii) If φ is the formula $\neg \psi$, $(D, E) \models \neg \psi\,(v)$ if and only if *not* $(D, E) \models \psi\,(v)$ (written $(D, E) \not\models \psi\,(v)$).

(iv) If φ is the formula $[\psi \Rightarrow \chi]$, $(D, E) \models [\psi \Rightarrow \chi]\,(v)$ if and only if $(D, E) \models \psi\,(v)$ *implies* $(D, E) \models \chi\,(v)$ (i.e. if and only if: if $(D, E) \models \psi\,(v)$, then also $(D, E) \models \chi\,(v)$; this is *material implication*, and is equivalent to: either $(D, E) \models \psi\,(v)$ is false, or $(D, E) \models \chi\,(v)$ is true).

(v) If φ is the formula $\exists x \psi(x)$, $(D, E) \models \exists x \psi(x)\,(v)$ if and only if there is some x-variant assignment v' of v for which $(D, E) \models \psi(x)\,(v')$. Here v' is an x-variant assignment of v if $v'(y) = v(y)$ for every variable $y \neq x$ (but $v'(x)$ can be any object of D, and in particular may differ from $v(x)$).

FIRST-ORDER LOGIC AND ITS USE IN SET THEORY

This notion of <u>variant assignments</u> allows us to express the intended meaning of the quantifier $\exists x$ reasonably concisely. It is at this point that we have ensured that our variables are indeed ranging over the domain D, since we have restricted our assignments to D, but then in the notion of an x-variant we have allowed any object of D to be assigned to x, so that no further restriction is made.

Note also here that we are following a common convention in (v) in writing $\exists x \psi(x)$ rather than $\exists x \psi$, in order to emphasize that the variable x is expected to occur free in the subformula $\psi(x)$ (though it need not; but the notion being explained would then be redundant). The distinction between (x) and (v) is then important, and should be easy enough for the reader to make.

2.4.2 Other defined symbols

As a check on the understanding of these definitions, the reader should note the following, which are simple consequences of the definitions in 2.2.8 and the definition 2.4.1 above:

(i) $(D, E) \models [\psi \wedge \chi] (v)$ if and only if *both* $(D, E) \models \psi (v)$ *and* $(D, E) \models \chi (v)$.
(ii) $(D, E) \models [\psi \vee \chi] (v)$ if and only if *either* $(D, E) \models \psi (v)$ *or* $(D, E) \models \chi (v)$ (or both).
(iii) $(D, E) \models [\psi \Leftrightarrow \chi] (v)$ if and only if either $(D, E) \models \psi (v)$ and $(D, E) \models \chi (v)$, or $(D, E) \not\models \psi (v)$ and $(D, E) \not\models \chi (v)$.
(iv) $(D, E) \models \forall x \psi(x) (v)$ if and only if for every x-variant assignment v' of v, $(D, E) \models \psi(x) (v')$.
(v) $(D, E) \models \exists^1 x \psi(x) (v)$ if and only if there is exactly one x-variant assignment v' of v such that $(D, E) \models \psi(x) (v')$.

Also we have the following:

2.4.3 Assignments for free variables only

If v and v' are two assignments in D such that for every variable x free in φ, $v(x) = v'(x)$, then $(D, E) \models \varphi (v)$ if and only if $(D, E) \models \varphi (v')$.

In view of this, if φ is a <u>sentence</u>, then we write $(D, E) \models \varphi$, and say φ is <u>true</u> or <u>valid</u> in (D, E), if and only if there is some assignment v in D such that $(D, E) \models \varphi (v)$ (equivalently if and only if for every assignment v in D, $(D, E) \models \varphi (v)$).

Also we say that φ is <u>logically valid</u> or <u>universally valid</u> if φ is valid in every structure (D, E).

If φ is not a sentence, but a formula with free variables, then we extend the notation by writing also $(D, E) \models \varphi$ if and only if for <u>every</u> assignment v in D, $(D, E) \models \varphi (v)$. Then if we write φ' for the <u>universal closure</u> of φ, then

---

"structure (D, E) satisfies wff. φ"

Notation: $(D, E) \models \varphi$ means $(D, E) \models \varphi(v)$ for <u>every</u> assignment v

★ Sentences contain no free vars. so every assignment for a sentence is null

} sentence p. 16

} ★

Theory ≡ set of formulas

22   INTERMEDIATE SET THEORY

$(D, E) \models \varphi$ if and only if $(D, E) \models \varphi'$. This is sometimes called the <u>universal</u> or <u>generality</u> interpretation of free variables.

Another extension of the notation is the following: we may write a list $\sigma_1, \sigma_2, \ldots, \sigma_n$ of objects of the domain D, to indicate the assignment v for which $v(x_1) = \sigma_1$, etc., and then we write $(D, E) \models \varphi\,(\sigma_1, \ldots, \sigma_n)$ (implying that the free variables of φ are the appropriate ones). See 2.5.1 for a use of this notation.

2.4.4 Semantic consequence

We say (D, E) is a <u>model</u> of φ if $(D, E) \models \varphi$ (as in 2.4.3 above); and if Γ is a set of formulas, we say (D, E) is a model of Γ, and write $(D, E) \models \Gamma$, if $(D, E) \models \varphi$ for every φ in Γ. Then we write $\Gamma \models \varphi$, and say φ is a <u>semantic consequence</u> of Γ, if every model of Γ is also a model of φ. (Note that this will be counted as true in the case when Γ has no models at all, i.e. is <u>inconsistent</u>.) Now the important fact referred to at the end of 2.3.3 is the following:

Theorem 2.4.5 *(Completeness theorem)* $\Gamma \vdash \varphi$ *if and only if* $\Gamma \models \varphi$.

In words: φ is a semantic consequence of Γ if and only if φ can be <u>deduced</u> from Γ. We shall not prove this here, but refer the reader to the references on first-order logic ([Sho67], [End72], [Men64], [Ham88]) for a proof.

2.4.6 Limitations of first-order logic

Although the completeness theorem is a strong theorem, it can be looked at as showing that first-order logic is actually quite weak. The first property to note is <u>compactness</u>. Because deductions and proofs are always finite sequences of formulas, any proof can only use a finite number of premises. In particular, if a set of formulas (a <u>theory</u>) is inconsistent, there will be a proof of this, and this proof can only use a finite subset of the formulas. Since the completeness theorem tells us that having a model is the same as being consistent, this can be stated as the following:

Theorem 2.4.7 *(Compactness theorem)* *A set Γ of formulas has a model if and only if every finite subset of Γ has a model.*

Two simple consequences of this are the following. First, no theory in a first-order language can have infinite models, all of which are isomorphic; indeed, if the theory is in a countable language (as LST is), then if it has an infinite model, it will have models of every infinite cardinality (which can't therefore all be isomorphic). Second, no theory in a first-order language can be such that in all of its models, some definable relation is always well-ordered (or well-founded), unless it is always finite of bounded length.

"Γ is consistent" ≡ "Γ has (at least one) model"
   inconsistent   no

FIRST-ORDER LOGIC AND ITS USE IN SET THEORY   23

We shall indicate proofs of these in exercises when we have introduced the appropriate notions; and we shall point out consequences for set theory in several places.

However there are also positive consequences. As a consequence of the completeness theorem, we can without affecting the results, take very varying attitudes to the formalization of mathematics in general and set theory in particular. We could insist on complete formality in the sense of allowing as mathematics, only completely formal proofs from formally stated axioms. Or we could disregard formal rigour altogether and use any means that come to hand, to convince ourselves or others that if a certain structure (such as the cumulative type structure) has certain first-order properties, then it has certain others. Then (provided our means of convincing ourselves are in fact correct), we shall come to the same first-order conclusions. Most mathematicians will take an attitude somewhere between these extremes; in particular there are a wide variety of more or less formal methods, all of which have been formally checked (on many occasions; they provide many useful exercises for students in logic). Having checked them, we know they are correct. And we have the assurance that we are not missing some correct methods of argument which have somehow been overlooked; the simple rules and axioms of first-order logic that we have presented are complete. (Of course that does not ensure that we have not missed some valid argument that we could have made, using methods that we knew well, but in fact overlooked at the time—most mathematicians can think of such occasions!)

All of this contrasts with the situation which arises when we ask, what first-order properties our intended structures actually have. We know from Gödel's incompleteness theorems (see 9.6.2) that for structures such as the natural numbers, or the cumulative type structure, no formal system of first-order axioms is complete. However many assumptions about these structures (axioms of arithmetic or set theory) we manage to convince ourselves of, there will always be further first-order statements independent of what we have so far.

2.4.8   Second-order logic

Second-order logic arises when we add to our language such quantifiers as: *for all subsets* (of some given set) or *for all functions* (from one given set to another). Both these quantifiers can be expressed quite simply within set theory; but if we ask about the consequences of second-order statements, it turns out that there is an important difference between the obvious intended meaning of a second-order quantifier (in the simple sense that *all subsets* or *all functions* should really mean *all*, not just those that we can define or represent in some model), and the equivalent statement in LST, which is first-order and when interpreted can only refer to those subsets or functions which happen to be in the model. This difference shows very clearly in the description we have given of the cumulative types: at each level, we want *all* subsets of the previous levels. But our first-order axioms can only capture an approximation to this idea. If we had an axiomatization of second-order consequence, with a completeness theorem on a par with the completeness theorem in 2.4.5, we might have means to answer all questions about the cumulative hierarchy; but of course there is no such axiomatization, and there cannot be. This follows from Gödel's incompleteness theorems, and really means that the study of second-order logic and the study of set theory (as we have described it) amount to the same thing. All we can do is to study first-order axioms for sets, and it could be likened to studying first-order *approximations* to second-order logic or set theory. We shall point this out as it happens at various points in our development.

2.5 Adding new terms

Our primitive language LST has only variables as terms, representing objects (sets). It is almost impossibly difficult to work in set theory without relaxing this and introducing many defined terms; in fact it is usual to be as liberal as possible, and allow any terms that we can, provided they do not lead to new conclusions which could not have been derived without them. This leads to the notion of a *conservative extension* of a theory: given a theory T in a language L, we say that a theory T' in a language L' which extends L, is a conservative extension of the theory T if (i) $T \subseteq T'$ and (ii) for any formula φ of the smaller language L, if $T' \vdash \varphi$ then already $T \vdash \varphi$.

For our purposes, L would be the primitive language LST, and L' would be LST with many added terms. T would be the theory whose axioms were the axioms of ZFC written in basic terms (without any new defined terms); T' would add axioms expressing the intended definitions of all the new defined terms. We want to ensure that T' is a conservative extension of T. Then in particular we can be sure that if T is consistent, so is T' (or conversely, that if T' is inconsistent, the fault already lay with T).

The following theorem gives sufficient conditions for all this:

Theorem 2.5.1 *Let T be a theory in a language with equality, and let $\varphi(x, x_1, \ldots, x_n)$ be a formula with just the free variables shown such that $T \vdash \forall x_1, \ldots, x_n \exists^1 x \varphi(x, x_1, \ldots, x_n)$. Let T' be an extension of T which is obtained by adding a new n-place function symbol f, and the axiom $\forall x_1, \ldots, x_n \varphi(f(x_1, \ldots, x_n), x_1, \ldots, x_n)$. Then T' is a conservative extension of T.*

Since this is a result in logic rather than in set theory, we only sketch a model-theoretic proof here, referring the interested reader to e.g. [Sho67] for

FIRST-ORDER LOGIC AND ITS USE IN SET THEORY

a more formal, complete proof.

Proof (Sketch) Suppose we are given a structure $\langle D, E \rangle$ which is a model of T (for simplicity we only treat the case where T is in the language LST). Then since $T \vdash \forall x_1, \ldots, x_n \exists^1 x \varphi(x, x_1, \ldots, x_n)$, we have $\langle D, E \rangle \models \forall x_1, \ldots, x_n \exists^1 x \varphi(x, x_1, \ldots, x_n)$. So given any n elements $d_1, \ldots, d_n$ from D there will be a unique element d in D for which $\langle D, E \rangle \models \varphi(x, x_1, \ldots, x_n)(d, d_1, \ldots, d_n)$. Let $F : D^n \to D$ be given by $F(d_1, \ldots, d_n) = d$; then $\langle D, E, F \rangle$ is a structure which is a model for the theory T'. If ψ is a formula of LST (i.e without the function symbol f), then $\langle D, E, F \rangle \models \psi$ if and only if $\langle D, E \rangle \models \psi$. So if $T \not\vdash \psi$, there will be a structure $\langle D, E \rangle$ which is a model of T in which ψ fails; the construction above then gives a structure $\langle D, E, F \rangle$ which is a model of T' in which ψ fails. So $T' \not\vdash \psi$. □

The terms we want to add will in fact be <u>abstraction terms</u> of the form $\{y \mid \theta\}$ for some formula θ, and the free variables of θ will be y and the variables $x_1, \ldots, x_n$ of the above proof; we shall assume that x does not appear in θ. The formula $\varphi(x, x_1, \ldots, x_n)$ will be $\forall y(y \in x \Leftrightarrow \theta)$. So for the application of 2.5.1 we shall need to prove $\forall x_1, \ldots, x_n \exists^1 x \forall y(y \in x \Leftrightarrow \theta)$. In every case, the axiom of extensionality will enable us to prove the uniqueness required, provided we can prove existence, i.e. provided we can prove $\forall x_1, \ldots, x_n \exists x \forall y(y \in x \Leftrightarrow \theta)$. So this is the form which most of our axioms will take.

We comment further on this when we come to use it in chapter 3.

2.5.2 Exercises

(1) Eliminating abstraction terms. Show that every formula is equivalent to a formula without abstraction terms, provided these have been introduced in a theory T in accord with the criterion in 2.5.1.
   [If f is introduced to T with the axiom

$$\forall x_1, \ldots, x_n \varphi(f(x_1, \ldots, x_n), x_1, \ldots, x_n)$$

and with

$$T \vdash \forall x_1, \ldots, x_n \exists^1 x \varphi(x, x_1, \ldots, x_n)$$

as above, then a formula $\psi(f)$ will be equivalent in T to

$$\exists y(\varphi(y, x_1, \ldots, x_n) \wedge \psi(y)).$$

Note that if we use abstraction terms, say s, t without first proving existence, there can be some doubt as to the meaning of e.g. $s \in t$; does it hold if t is really a proper class, when s satisfies the condition defining t? Presumably for $s \in t$ to hold one will require that s exist (i.e. that s be a set). However it is clear that one can still write equivalent formulas

without these terms, whichever interpretation one wants; it is just that there will be different formulas for the different interpretations.]
(2) Show that every formula is equivalent to a formula in *prenex form*, that is, all the quantifiers appear on the left, so that the formula has the form

$$Q_1 x_1 \ldots Q_k x_k \varphi$$

where each Q_i is either $\forall$ or $\exists$, and φ has no quantifiers.

[First remove abstraction terms, then bring quantifiers to the left using equivalences such as

$$\neg \forall x \varphi \Leftrightarrow \exists x (\neg \varphi)$$
$$\neg \exists x \varphi \Leftrightarrow \forall x (\neg \varphi)$$

and

$$(\forall x \varphi \Rightarrow \psi) \Leftrightarrow \exists x (\varphi \Rightarrow \psi)$$
$$(\psi \Rightarrow \forall x \varphi) \Leftrightarrow \forall x (\psi \Rightarrow \varphi)$$
$$(\exists x \varphi \Rightarrow \psi) \Leftrightarrow \forall x (\varphi \Rightarrow \psi)$$
$$(\psi \Rightarrow \exists x \varphi) \Leftrightarrow \exists x (\psi \Rightarrow \varphi)$$

provided x is not free in ψ. (Think of $\varphi \Rightarrow \psi$ as $\neg \varphi \vee \psi$ if these seem strange.) It may be necessary to change bound variables in order to complete this.]

3

The axioms of set theory

3.1 The Zermelo–Fraenkel axioms

We give here a set of axioms which have been somewhat modified since [Zer08]; in current usage, the Zermelo–Fraenkel axioms, ZFC, are taken to be anything equivalent to the axioms presented here, and ZF refers to anything equivalent to these axioms minus the axiom of choice. We give the axioms first, then further comments on their justification in 3.2.

3.1.1 A1 The axiom of extensionality

We have put the relation of equality into our logic, in 2.3, together with its standard interpretation in 2.4. So the first axiom, the axiom of extensionality, which is to give force to our intention that sets with the same members are the same set, can be stated as:

$$\forall x[x \in a \Leftrightarrow x \in b] \Rightarrow a = b$$

"If a and b have the same members, they are the same set." Here a and b are left as free variables; but we are assuming the generality interpretation of free variables, and the axiom means exactly the same as if we had preceded it with the universal quantifiers $\forall a \forall b$. We leave them out only for readability.

This axiom already enforces the decision (mentioned in 1.4) to accept only the empty set as an urelement. Any other urelement would also have no members (i.e. exactly the same members as the empty set), and so would be equal to the empty set by this axiom. If we did want to allow other urelements besides the empty set, this axiom would have to be weakened (see exercise 3.2.3(1)).

It also enforces the decision to ignore any question about repetition of membership; an element is either a member, or not a member. It makes no difference if it is counted once or several times.

Let us note here one major use we shall make of this axiom: most of our remaining axioms will be there to justify the use of some abstraction term

$\{y \mid \theta\}$. For this, as noted in 2.5, we must prove

$$\forall x_1, \ldots, x_n \exists^1 x \forall y (y \in x \Leftrightarrow \theta),$$

where $x_1, \ldots, x_n$ are all the free variables of θ apart from y, and x does not occur in θ. Omitting the initial universal quantifiers and expanding the definition of $\exists^1$, this is

$$\exists u \forall x (\forall y (y \in x \Leftrightarrow \theta) \Leftrightarrow u = x)$$

which is equivalent to the two formulas

$$\exists x \forall y (y \in x \Leftrightarrow \theta) \quad \text{and} \quad \forall y (y \in x \Leftrightarrow \theta) \wedge \forall y (y \in u \Leftrightarrow \theta) \Rightarrow u = x.$$

The second of these is an immediate consequence of the axiom of extensionality, since

$$\forall y (y \in x \Leftrightarrow \theta) \wedge \forall y (y \in u \Leftrightarrow \theta) \Rightarrow \forall y (y \in x \Leftrightarrow y \in u).$$

Hence to justify the introduction of the abstraction term $\{y \mid \theta\}$ we need to prove (or take as an axiom) the formula $\exists x \forall y (y \in x \Leftrightarrow \theta)$; the rest follows by the axiom of extensionality and 2.5.1.

Note that the commonest abbreviations for abstraction terms, some of which are used in the axioms below, are listed in chapter 10.

3.1.2   A2 The null-set axiom

or empty set axiom

$$\exists x \forall y (y \in x \Leftrightarrow y \neq y)$$

"There is a set with no members". This is the simplest axiom which justifies an abstraction term, in this case $\{y \mid y \neq y\}$, which we shall abbreviate as usual as $\emptyset$ (since $y \neq y$ is always false, so that this set has no members).

3.1.3   A3 The pair-set axiom

$$\exists x \forall y (y \in x \Leftrightarrow y = a \vee y = b)$$

"There is a set whose members are just a and b". This justifies the term $\{y \mid y = a \vee y = b\}$, which we shall abbreviate as $\{a, b\}$ as in usual mathematical usage for the unordered set with the members a and b only.

3.1.4   A4 The sum-set axiom

or union axiom

$$\exists x \forall y [y \in x \Leftrightarrow \exists z (y \in z \wedge z \in a)]$$

THE AXIOMS OF SET THEORY

"There is a set whose members are exactly the members of the members of a". This justifies the term $\{y \mid \exists z(y \in z \land z \in a)\}$, which we abbreviate as $\bigcup a$, the *union* or *sum-set* of the set a.

3.1.5 A5 The power-set axiom

$$\exists x \forall y(y \in x \Leftrightarrow y \subseteq a)$$

"There is a set whose members are exactly the subsets of a". Here $y \subseteq a$ abbreviates $\forall z(z \in y \Rightarrow z \in a)$, as usual, and this axiom justifies the *power-set* $\mathcal{P}(a)$ or $\{y \mid y \subseteq a\}$.

3.1.6 A6 The subset axiom

or axiom of separation (in German *Aussonderungsaxiom*)

$$\exists x \forall y[y \in x \Leftrightarrow y \in a \land \varphi(y)]$$

"For any formula $\varphi(y)$ there is a set whose members are exactly those members y of a which satisfy $\varphi(y)$". This is an axiom schema; we get one axiom for each formula $\varphi(y)$, with the condition that the variable x should not appear in $\varphi(y)$. It justifies the set $\{y \mid y \in a \land \varphi(y)\}$, which is a sub-set of a; hence the name. The second name is from the idea that the new subset separates those members of a which satisfy $\varphi(y)$ from those which don't. Yet another name used for it is the *restricted comprehension principle*, since it corresponds to the general comprehension principle with the formulas restricted to those of the form $y \in a \land \varphi(y)$. This is, of course, a very strong restriction.

3.1.7 A7 The replacement axiom

$$\forall z \forall u \forall v[\psi(z,u) \land \psi(z,v) \Rightarrow u = v] \Rightarrow \exists x \forall y[y \in x \Leftrightarrow \exists z(z \in a \land \psi(z,y)]$$

"If $\psi(z,y)$ is functional [see below], then there is a set which is the image of set a under ψ". Again this is an axiom scheme, giving one axiom for each formula $\psi(z,y)$, and again the variable x must not appear in $\psi(z,y)$. It justifies the set $\{y \mid \exists z(z \in a \land \psi(z,y)\}$, *provided* that the formula $\psi(z,y)$ satisfies $\forall z \forall u \forall v[\psi(z,u) \land \psi(z,v) \Rightarrow u = v]$, in other words provided that for each z there is at most one set y for which $\psi(z,y)$ holds (such a formula then expresses a predicate which is called *functional*; it describes a partial function). If this holds, then the set $\{y \mid \exists z(z \in a \land \psi(z,y)\}$ will have at most one member y for each member z of a. We say that the elements y *replace* the elements z from a; hence the name.

If we allow an extension of the language, and introduce function letters such as F to denote the function being described by ψ (so that $\psi(z,y) \Leftrightarrow F(z) = y$), then the set $\{y \mid \exists z(z \in a \wedge \psi(z,y)\}$ will be $\{F(z) \mid z \in a\}$ (that is, the *image* of a under F). Such extensions of the language are presented in chapter 9. When we come to use the replacement axiom in 6.6 we shall introduce abbreviations which allow us to use this sort of simplification, without in fact extending the primitive language.

3.1.8   A8 The axiom of foundation

$$\exists x(x \in a) \Rightarrow \exists x(x \in a \wedge x \cap a = \emptyset)$$

"Every non-empty set has a member which is disjoint from it". This axiom is not justifying anything; it is rather restricting attention to sets which occur in the cumulative type structure, since these will all have the property required. To see this, suppose that a has a member. Then if we start ascending the cumulative types, until we come to the first level at which a has a member, say x, then $x \cap a$ must be empty (since the members of x, if any, will all be at lower levels than x itself, and therefore cannot also be members of a). Another justification is the derivation of it in 9.5 from the axioms given there for ranks directly.

Note that the axiom could be written with $x \cap a = \emptyset$ replaced by $\forall z \neg (z \in x \wedge z \in a)$. That way, it will not be dependent on the presence of any other axioms (such as the axioms needed to justify $x \cap a$ or $\emptyset$), before it has its correct meaning. This is only important if we are investigating the independence of the axioms, as in exercise 3.2.3(2), or if we want to take weak subsystems of the axioms; since we are more interested in the ease of understanding the axioms, we give the more readable form.

3.1.9   A9 The axiom of infinity

$$\exists w[\emptyset \in w \wedge \forall x(x \in w \Rightarrow x \cup \{x\} \in w)]$$

"There is a set with $\emptyset$ as a member, which is closed under the function $x \mapsto x \cup \{x\}$". This is needed to distinguish set theory, which was introduced to deal with infinite objects, from arithmetic, which is all that is needed if only finite objects are being considered. The set w which is postulated to exist, cannot be finite, since (as we shall note in exercise 3.2.3(3)), the operation of forming $x \cup \{x\}$ must always give a new member. As with the axiom of foundation, we could (and for some purposes should) write out the axiom in a way that does not rely on other axioms to justify $x \cup \{x\}$; but for ease of readability, we give the form above.

THE AXIOMS OF SET THEORY

3.1.10   A10 The axiom of choice

$$\forall x \forall y (x \in a \land y \in a \land x \neq y \Rightarrow x \cap y = \emptyset) \land \forall x(x \in a \Rightarrow x \neq \emptyset)$$
$$\Rightarrow \exists c \forall x [x \in a \Rightarrow \exists u (c \cap x = \{u\})]$$

"Every set of disjointed non-empty sets has a choice set". We shall say a lot more about this axiom in chapter 7; for now, let us just note that the two hypotheses on a say that a has only non-empty members which are pairwise disjoint. The conclusion is that there is a choice set c which has exactly one member in common with each member of a; we can say that c *chooses* those members from the members of a. It can be thought of as an attempt to enforce the intention that at every level of the cumulative type structure, *every* subset is a set at the next level. So since all the members of a are at lower levels, so will be all the members of the members, and *all* subsets formed from them will exist at least at the level where a exists. Amongst them will be many that select one member from each member of a.

3.2   Arguments for these axioms

On the basis of the cumulative type structure, we want to argue that the axioms given above are obviously true. This is intended in exactly the sense of the definition of truth (or satisfaction) as given in 2.4, with the structure taken to be the cumulative types, together with the membership relation.

The axiom of extensionality is true since, if we have two sets a and b such that $\forall x(x \in a \Leftrightarrow x \in b)$ is true (with set a assigned to the variable a and set b to the variable b; this is a very convenient way of thinking of sets, as being labelled by a name which happens to be just the same as the variable we are going to assign to them), then a and b must have exactly the same members, they must exist at exactly the same level, and we have agreed to regard them as identical. So $a = b$ is also true (at the same assignment). But note that we are implicitly agreeing that there are not any other objects than sets; as noted when stating the axiom, this form of the axiom rules out any urelements other than the empty set.

The null-set axiom will be true since there is in fact a null-set or empty set amongst our sets in the cumulative type structure (existing at the first level, if not taken as an urelement). If we denote this by $\emptyset$, and let v be an assignment with $v(x) = \emptyset$, then since $y \neq y$ is always false, $\forall y(y \in x \Leftrightarrow y \neq y)$ will be satisfied at the assignment v. So the null-set axiom will be satisfied.

The pair-set axiom does require that we have no last level, if it is to be true: $\forall y(y \in x \Leftrightarrow y = a \lor y = b)$ will only be satisfied if x is assigned to the pair-set of a and b (where again we assume set a is assigned to variable a and set b to variable b—we shall make this assumption without mention in future). But this pair-set will exist at any level after a level at which both a and b exist, so, assuming there is no last level, the pair-set axiom will be true.

The sum-set axiom will hold even if there is a last level; if a exists at some level, then so do all its members and all their members, and the collection of all these members of members of a (i.e. $\bigcup a$) will certainly exist at the same level as a. (It may exist at the level before a if there is such a level; this is illustrated in exercise 6.5.8(1).) Then $\forall y[y \in x \Leftrightarrow \exists z(y \in z \land z \in a)]$ will be satisfied if and only if x is assigned to $\bigcup a$; since we can make this assignment, the sum-set axiom will be satisfied.

The power-set axiom, like the pair-set axiom, requires that there be no last level; but then the argument for it is the same. $\forall y(y \in x \Leftrightarrow y \subseteq a)$ will be satisfied if and only if x is assigned to $\mathcal{P}(a)$, and since $\mathcal{P}(a)$ will exist at the level after the one at which a exists, we can make this assignment, and the power-set axiom will be satisfied.

The subset axiom doesn't need any further levels; like the sum-set axiom, it would still be true if there were a last level. $\forall y[y \in x \Leftrightarrow y \in a \land \varphi(y)]$ will be satisfied if and only if x is assigned to the subset of a consisting of those members of a for which $\varphi(y)$ is satisfied. But this set, $\{y \mid y \in a \land \varphi(y)\}$, will exist at or before the level at which a exists, and so will be available for x to be assigned to; and the subset axiom will be satisfied.

The argument for the replacement axiom is less simple. It is the strongest statement we make about the levels having no conceivable end, and can be summarized as follows: suppose that certain levels are correlated with the members of a set. Then that collection of levels can be considered as completed, in the sense that there must be a further level beyond (or above) all of them.

So if we take an instance of the replacement axiom, and if $\psi(x, y)$ is the formula assumed to express a functional predicate, then the level at which y occurs is being correlated to the element x; and if we then consider only those elements x which are members of the set a, we have a collection of levels correlated with the members of a set, as described above. At any level which lies beyond all of them, the required set $\{y \mid \exists z(z \in a \land \psi(z, y))\}$ will exist, and so the instance of the replacement axiom will hold.

There is an alternative justification of the replacement axiom which goes back to Cantor's distinction between consistent and inconsistent manifolds, and his explanation that "Two equivalent manifolds are either both sets or both inconsistent" (see 1.1). The collection $\{y \mid \exists z(z \in a \land \psi(z, y))\}$ is easily seen to be equivalent to a or a subset of a, and so should be a set (since a is). (The notion of *equivalent* used here is in 4.1.1 if it is new to the reader.) This sort of argument is sometimes referred to as the "limitation of size theory": a collection is a set if its size is limited (in some appropriate way). See [Hal84].

The axioms of foundation and choice were given their arguments above, and we shall not say more about them now.

THE AXIOMS OF SET THEORY 33

3.2.1 Classification of these axioms

It is worthwhile to think of these axioms in the following way: first, there are axioms which restrict attention to the cumulative notion we have described; these are the axioms of extensionality and foundation. The remaining axioms then demand that there are enough sets (or enough levels). The null-set and pair-set axioms are the simplest, and the union axiom and the power-set axiom then enforce the existence of definable sets which are rather easy to conceive of. The subset axiom is a further attempt to get at *all* the sets of the cumulative hierarchy; but it is known to be insufficient, and the axiom of choice is also needed.

Finally the axioms of infinity and replacement say something about the extent of the hierarchy; they have been called "axioms of extent", and further axioms of extent have been proposed (we shall describe the first steps in this direction in 9.4). At this point let us notice that the finite levels of the cumulative hierarchy form what are also called the *hereditarily finite sets* (see exercise 3.2.3(4)), and all the axioms apart from the axiom of infinity are true within the hereditarily finite sets. Adding the axiom of infinity forces consideration of levels beyond all the finite levels (this is shown in the exercise), and it will be true just when there is such an infinite level. Then the replacement axiom will require that the levels continue for a very long way beyond the first infinite level; and the combination of the power-set axiom with the axioms of infinity and replacement gives a very rich structure of levels.

3.2.2 Dependencies

These axioms are not independent; dependencies among them are set out in exercise 3.2.3(2). In particular, the replacement axiom scheme implies the subset axiom scheme, and in conjunction with the power-set axiom it implies the pair-set axiom. The null-set axiom follows from the subset axiom. The main reason for considering all these axioms, and not just a minimal subset which suffices to prove the rest, is that there is considerable interest in weaker subsystems of set theory, which omit some of the stronger axioms but not all of their simpler consequences. So we may want to omit the replacement axiom, but we would probably not want to also omit the pair-set axiom. Many other weakenings have been considered: a few are mentioned in later chapters.

3.2.3 Exercises

(1) Weak forms of extensionality. The simplest weakening of extensionality was used by Fraenkel and Mostowski in giving independence proofs (mentioned in chapter 8). This just says that extensionality applies to only

[handwritten note: Only individual is the empty set. Allowing distinct multiple empty sets means allowing distinct multiple individuals]

sets with members: *..., and not the empty set*

$$\exists x(x \in a) \Rightarrow [\forall x(x \in a \Leftrightarrow x \in b) \Rightarrow a = b]$$

Show that this suffices for any application justifying the introduction of an abstraction term, provided that the term is not empty.

A more general change which also allows for a unique empty set is to use a language with two <u>sorts</u> of variables, one sort for sets (to which extensionality will apply) and the other for more general objects (usually taken to include sets but also other urelements). If $a, b, \ldots$ are variables for sets, and $x, y, \ldots$ are variables for more general objects including sets, then extensionality can be taken exactly as in 3.1.1. It would be normal then to add an axiom to restrict membership so that any object which has members has to be a set:

$$\exists x(x \in y) \Rightarrow \exists a(y = a).$$

How should the null-set axiom be written in this context?

Show that if the other set existence axioms are written correctly, we can still show e.g. that if sets A and B are disjoint, and so are sets C and D, then $A \cap B = C \cap D$.

Write axioms in this context to say that the collection of urelements is not a set; or to say that it is a set.

(2) Show that the replacement axiom scheme implies the subset axiom scheme.

[Take $\psi(z, y)$ as $y = z \wedge \varphi(z)$.]

Show that the null-set axiom follows from the subset axiom.

[Note that it is usually taken as an axiom of logic that the universe is not empty; this is implied by 2.3.1(5), from which we can deduce $\forall x \varphi \Rightarrow \exists x \varphi$, and by the requirement in 2.4 that the domain of a structure be non-empty. Without this the null-set axiom would not follow from the subset axiom; some statement of the existence of something is needed first. The axiom of infinity is the only other direct statement of existence in our axioms.]

Show that the pair-set axiom follows from the other axioms: show that the second power-set (the power-set of the power-set) of the empty set has two members, and then use the replacement axiom to prove the pair-set axiom.

(3) Consequences of the axiom of foundation. Show from the axiom of foundation and the pair-set axiom that $x \notin x$ and that $x \in y \Rightarrow y \notin x$.

[Apply the axiom of foundation to the sets $\{x\}$ and $\{x, y\}$, and get a contradiction if $x \in x$ or if $x \in y \wedge y \in x$.]

Hence show that $x \neq x \cup \{x\}$ and that $x \cup \{x\} = y \cup \{y\} \Rightarrow x = y$.

(4) The hereditarily finite sets. These are given by the recursive definition

(i) ∅ is a hereditarily finite set;
(ii) if $x_1, x_2, \ldots, x_n$ are all hereditarily finite sets, so is $\{x_1, x_2, \ldots, x_n\}$;
(iii) these are all the hereditarily finite sets.

Show that the collection of all hereditarily finite sets forms a model for all of the axioms except the axiom of infinity.

[If we let H be the collection of all hereditarily finite sets, then the structure we are considering is the pair $\langle H, \in \rangle$, or really $\langle H, E \rangle$ where E is the membership relation restricted to H. Extensionality is automatic if we take any collection of sets with the membership relation, provided the collection is closed under membership as H is (is *transitive*, as we shall use this term later in 6.1.8). The axioms of pair-set, sum-set, power-set, subset and replacement, and the axiom of choice all hold because the required members of the set needed can be listed in a finite list, and so form a hereditarily finite set by (ii). The axiom of foundation can be shown to hold, and the axiom of infinity to fail, by defining *levels* for these sets: say that ∅ has level 0, and if $x_1, x_2, \ldots, x_n$ have levels $l_1, l_2, \ldots, l_n$ respectively, then $\{x_1, x_2, \ldots, x_n\}$ has level $\max\{l_1, l_2, \ldots, l_n\} + 1$. Then since members must have lower levels than the set they are members of, a member of a of minimal level will be disjoint from a and the axiom of foundation must hold. And if $x \in w$ has *maximal* level (for members of w) then $x \cup \{x\} \notin w$ and the axiom of infinity must fail; note the importance of w being finite for this maximal level to exist.]

4

Cardinals

We shall give in this chapter some of the development of the theory of cardinals. This was one of the first parts of set theory developed by Cantor, and we shall give it here in a naive form very much as it was originally given. We shall later want to examine how this work can be formalized, and also the extent to which the axioms we gave are sufficient for it.

4.1 Countable sets

Definition 4.1.1 Two sets A and B are *similar* (or *equinumerous* or *numerically equivalent*), written $A \sim B$, if there is a bijection from A to B. ("Equivalent" will mean "numerically equivalent" unless otherwise stated.)

Proposition 4.1.2 *The relation $A \sim B$ is an equivalence relation on the class of sets.*

Proof For any sets A, B and C, we have:

(i) $A \sim A$ via the identity map;
(ii) $A \sim B$ implies $B \sim A$ since the inverse of a bijection is a bijection;
(iii) $A \sim B$ and $B \sim C$ implies $A \sim C$ since the composition map will serve.

□

Definition 4.1.3 Finite and infinite sets: a set A is called *finite* if it is empty or if there is a natural number n and a bijection from $\{1, 2, \ldots, n\}$ to A. A set is called *infinite* if it is not finite.

Note that these notions are taken to be naive notions, and later we shall want to give formal equivalents and prove, for example, that the set $\mathbb{N}$ is infinite, i.e. that it is impossible to find an $n \in \mathbb{N}$ and a bijection $\mathbb{N} \to \{1, 2, \ldots, n\}$ (a fact which should be obvious; the point of proving it

4.1.4 Examples

(i) $\{0, 1, 2, \ldots, n, \ldots\} \sim \{1, 2, 3, \ldots, n, \ldots\}$ via the bijection $n \mapsto n+1$, $n \in \mathbb{N}$.
(ii) If $\mathbb{E}$ is the set of even numbers, $\mathbb{N} \sim \mathbb{E}$ via the bijection $n \mapsto 2n$, $n \in \mathbb{N}$.
(iii) $\mathbb{R} \sim \mathbb{R}^+$ via the bijection $x \mapsto e^x$, $x \in \mathbb{R}$.
(iv) $\mathbb{R} \times \mathbb{R} \sim \mathbb{C}$ via the bijection $(x, y) \mapsto x + iy$, $(x, y) \in \mathbb{R} \times \mathbb{R}$.
(v) $[0, 1] \sim [1, 3]$ via the bijection $x \mapsto 2x + 1$. (Here $[0, 1]$ is the closed unit interval of the real line.)
(vi) $(-\frac{\pi}{2}, \frac{\pi}{2}) \sim \mathbb{R}$ via the bijection $x \mapsto \tan x$. (Here $(-\frac{\pi}{2}, \frac{\pi}{2})$ is the open interval of the real line.)

A significant point from these examples is that an infinite set can be equinumerous with a proper subset of itself, which is impossible for finite sets. In fact we shall show, using the axiom of choice, that every infinite set has this property, and this was in fact taken as a definition of "infinite" by Dedekind:

Definition 4.1.5 A set is called *Dedekind infinite* if it is equinumerous with one of its proper subsets.

Thus $\mathbb{N}$ is Dedekind infinite, and as $\mathbb{N} \subset \mathbb{Z} \subset \mathbb{Q} \subset \mathbb{R} \subset \mathbb{C}$, these sets are also Dedekind infinite. But the axiom of choice is needed to show that every infinite set is Dedekind infinite (see 7.4.1).

Is there a *smallest* infinite set? Again, the axiom of choice is needed to prove that in a clear sense there is; but we can give the definitions now:

Definition 4.1.6 An infinite set A is said to be *denumerable* (or sometimes *enumerable*) if $A \sim \mathbb{N}$.

A set is said to be *countable* if it is either finite or denumerable.

So a set is countable if it can be written as a sequence $\{a_1, a_2, \ldots, a_n, \ldots\}$; if this can be done without repetitions, it is denumerable (via the bijection $n \mapsto a_n$).

Warning We have given a common usage; but many authors have used *denumerable* in the sense we have given for *countable*. The only safe way is to say that a set is *denumerably infinite* (or *countably infinite*), for what we have called denumerable above.

4.1.7 Examples

(i) Every finite set is countable.

CARDINALS

(ii) $\mathbb{N}$ is countable.
(iii) Each subset of $\mathbb{N}$ is countable.
(iv) If A is countable and $A \sim B$, then B is countable.

For proving countability, one needs to construct a suitable bijection. This can be made easier by using the notion of *dominance*:

Definition 4.1.8 Set A is *dominated* by set B (in symbols $A \preccurlyeq B$), if there is an injection from A into B. A is *strictly dominated* by B (in symbols $A \prec B$), if $A \preccurlyeq B$ and A is not equinumerous with B.

4.1.9 Examples

(i) A is finite implies $A \prec \mathbb{N}$.
(ii) $A \sim B$ implies $A \preccurlyeq B$; but the converse need not be true (e.g. by (i)).
(iii) $A \preccurlyeq A$ (reflexivity).
(iv) $A \preccurlyeq B$ and $B \preccurlyeq C$ implies $A \preccurlyeq C$ (transitivity).
(v) $A \subseteq B$ implies $A \preccurlyeq B$; but the converse need not be true.
(vi) If $A \preccurlyeq B$ and $A \sim C$ then $C \preccurlyeq B$.

Proposition 4.1.10 *(Fundamental theorem for demonstrating countability)*
A is countable if and only if $A \preccurlyeq \mathbb{N}$; in other words if and only if there is an injection $A \to \mathbb{N}$.

Proof Necessity is immediate from the definitions of finite and denumerable.
Sufficiency. Suppose $g : A \to \mathbb{N}$ is an injection; then $g(A)$ (i.e. $\{g(a) \mid a \in A\}$) is a subset of $\mathbb{N}$ and so countable. But $g : A \to g(A)$ is a bijection. □

Corollary 4.1.11 *A non-empty set A is countable if and only if there is a surjection from $\mathbb{N}$ onto A.*

Proof Given an injection $f : A \to \mathbb{N}$ with $A \neq \emptyset$, let $a_0 \in A$ be any particular member, and define $g : \mathbb{N} \to A$ by

$$g(n) = \begin{cases} f^{-1}(n) & \text{if } n \in f(A); \\ a_0 & \text{otherwise.} \end{cases}$$

Clearly g is a surjection onto A. Conversely, if $g : \mathbb{N} \to A$ is a surjection, we get an injection $f : A \to \mathbb{N}$ by taking $f(a)$ as the smallest $n \in \mathbb{N}$ such that $g(n) = a$. □

The direct construction of injections or surjections is often difficult, and many properties of $\mathbb{N}$ are useful in this. We used the "least element" property in the last proof; another important property is that $\mathbb{N}$ has "pairing functions", such as those given by using the unique decomposition of natural numbers into products of powers of primes. This is used in the following:

Proposition 4.1.12 *The countable union of countable sets is countable.*

Proof Let $(A_n)_{n \in \mathbb{N}}$ be any countable collection of sets; and we assume that each of the sets A_n is itself also countable. Then we must show that the union $A = \bigcup_{n \in \mathbb{N}} A_n$ is countable. So we may assume no A_n is empty, and using 4.1.11, we assume given a surjection $f_n : \mathbb{N} \to A_n$ for each $n \in \mathbb{N}$. (We shall note later in 7.4.2 that this is a naive use of the axiom of choice.) Then we get a surjection $f : \mathbb{N} \to A$ by defining

$$\begin{cases} f(0) = f_0(0); \\ f(n) = f_k(l) \end{cases} \text{where } 0 \neq n = 2^k.3^l.m \text{ and } m \text{ is not divisible by 2 or 3.}$$

□

An alternative proof is provided by considering the following diagram:

$$\begin{array}{l} A_1 = \{a_{11} \quad a_{12} \to a_{13} \quad \ldots\} \\ \quad \downarrow \nearrow \swarrow \nearrow \\ A_2 = \{a_{21} \quad a_{22} \quad a_{23} \quad \ldots\} \\ \quad \swarrow \nearrow \\ A_3 = \{a_{31} \quad a_{32} \quad a_{33} \quad \ldots\} \\ \quad \downarrow \nearrow \\ \quad \vdots \end{array}$$

which produces the listing

$$a_{11}, a_{21}, a_{12}, a_{13}, a_{22}, a_{31}, a_{41}, a_{32}, a_{23}, a_{14}, a_{15}, a_{24}, a_{33}, \ldots$$

to show that the union $A = \bigcup_{n \in \mathbb{N}} A_n$ is countable.

Proposition 4.1.13 *The set $\mathbb{Q}$ of all rational numbers is countable.*

Proof Let $\mathbb{Q}^+$ and $\mathbb{Q}^-$ be the positive and negative rational numbers, so that $\mathbb{Q} = \mathbb{Q}^+ \cup \{0\} \cup \mathbb{Q}^-$, and construct an injection $f : \mathbb{Q}^+ \to \mathbb{N}$ by:

$$f(x) = 2^p \times 3^q, \quad p \text{ and } q \text{ positive integers with } (p,q) = 1 \text{ and } x = \frac{p}{q}.$$

(An alternative injection can be given by:

$$f(\frac{p}{q}) = \frac{1}{2}(p+q-1)(p+q-2) + q;$$

it is instructive to compare the listing of $\mathbb{Q}^+$ implied by this last injection with the diagram giving the alternative proof of 4.1.12.)

Clearly $\mathbb{Q}^+ \sim \mathbb{Q}^-$, and the proof is completed by 4.1.12. □

Corollary 4.1.14 *The set of all rational points in $\mathbb{R}^n$ is countable.*

CARDINALS 41

Proof By 4.1.12 and 4.1.13, it follows that $\mathbb{Q} \times \mathbb{Q}$ is countable, and similarly $\mathbb{Q} \times \mathbb{Q} \times \mathbb{Q}$, etc. □

4.1.15 Exercises

(1) Show that the following sets are countable:

   (i) the set of all finite sequences from $\mathbb{N}$;
   (ii) the set of all finite subsets of $\mathbb{N}$;
   (iii) the set of all complex roots of unity;
   (iv) the set of all polynomials in one variable with integral coefficients;
   (v) the set of all algebraic numbers;
   (vi) the set of all formulas of a countable language (where a language is countable if its set of symbols is countable).

(2) Show that a set E is countable if and only if there is a bijection $f : E \to E$, such that the only subsets F of E such that f (when restricted to F) is a bijection from F to F, are $F = \emptyset$ and $F = E$.

   [Countable includes finite here. Take an element $x_0 \in E$, and consider the set $F = \{f^k(x_0) \mid k \in \mathbb{Z}\}$; show that f will be a bijection when restricted to F. For the converse, note that the mapping $k \mapsto k+1$ on $\mathbb{Z}$ has the property required.]

4.2 Uncountable sets

Cantor's major discovery was that not all sets are countable, so that there is more than one *size* of infinite set. This was a major upset for the theology of his day (which took infinity as synonymous with deity, and so insisted that there was only one infinity); and it was also quite revolutionary for the mathematics of his day. It is our first major result of set theory.

Definition 4.2.1 A set which is not countable is called *uncountable* or *nondenumerable*.

Theorem 4.2.2 *The set of all real numbers is uncountable.*

Proof We shall show that the open interval $(0, 1)$ is uncountable; it is easy to see that $(0, 1) \sim \mathbb{R}$ by methods as in examples 4.1.4.

Any real number in $(0, 1)$ can be written as an infinite decimal of the form $0.d_1 d_2 d_3 \ldots$, where each d_i is a digit, $0 \leq d_i \leq 9$. Now suppose that $(0, 1)$ were denumerable, say as $\{a_1, a_2, \ldots, a_n, \ldots\}$. Let

$$a_1 = 0.d_{11}d_{12}d_{13}\ldots$$
$$a_2 = 0.d_{21}d_{22}d_{23}\ldots$$
$$a_3 = 0.d_{31}d_{32}d_{33}\ldots$$
$$\vdots$$
$$a_n = 0.d_{n1}d_{n2}d_{n3}\ldots$$
$$\vdots$$

where each d_{ij} is a digit.

Now construct a real number

$$a^* = 0.c_1c_2c_3\ldots$$

where
$$c_i = 1 \quad \text{if} \quad d_{ii} \neq 1, \quad \text{and}$$
$$c_i = 3 \quad \text{if} \quad d_{ii} = 1 \quad \text{for each } i = 1, 2, 3, \ldots$$

Then since $c_i \neq d_{ii}$, we have $a^* \neq a_i$ for each $i = 1, 2, 3\ldots$; yet $a^* \in (0,1)$. This contradicts the assumption that the list $\{a_1, a_2, a_3, \ldots\}$ enumerates all of $(0,1)$. Since this list is quite arbitrary, we have shown that there cannot be such a list which enumerates $(0,1)$; so $(0,1)$ cannot be denumerable. □

Note that there is nothing very special about 1 and 3 in the above proof; but since many rational numbers do have two decimal expansions, one ending in 0 recurring and the other in 9 recurring (as $\frac{1}{2}$ is both $0.4999\ldots$ and $0.5000\ldots$), we have carefully avoided 0 and 9.

The method of proof used here came to be known as "Cantor's diagonal method", since it can be pictured as taking the array of decimal expansions given by the assumed enumeration, and, treating it as an infinite rectangle, changing all the digits down the main diagonal to get a new number, a^*, not already in the array.

It was refined by Cantor into the following, usually known as Cantor's theorem:

Theorem 4.2.3 *For any set A there is no surjection from A onto $\mathcal{P}(A)$.*

Proof Let $f : A \to \mathcal{P}(A)$ be any map; we show it is not surjective by constructing a subset of A not in its range. Note that since for each $x \in A$, $f(x) \subseteq A$, it makes sense to ask whether $x \in f(x)$ or not. So consider

$$T = \{x \in A \mid x \notin f(x)\}.$$

CARDINALS

Clearly $T \in \mathcal{P}(A)$; but suppose that $T = f(y)$ for some $y \in A$. Then $y \in f(y)$ implies $y \notin T$, and $y \notin f(y)$ implies $y \in T$, both by the definition of T. Both of these contradict $T = f(y)$, and so T is not in the range of f; i.e. f is not surjective. □

Since it is easy to construct an injection from A to $\mathcal{P}(A)$, e.g. $a \in A \mapsto \{a\} \in \mathcal{P}(A)$, we have:

Corollary 4.2.4 *For every set A, $A \prec \mathcal{P}(A)$.*

From this result we can now deduce Cantor's paradox, as noted in 1.3.4: if the collection of all sets were a set, A say, it would contain all of its subsets as members and so contradict 4.2.3.

After Cantor's theorem, the most important result of this stage of set theory is the Schröder–Bernstein theorem. This gives us bijections from two injections, and is a very useful way to get bijections:

Theorem 4.2.5 *For any sets A and B, if $A \preccurlyeq B$ and $B \preccurlyeq A$, then $A \sim B$.*

We give two proofs; the first uses a lemma due to Banach:

Lemma 4.2.6 *If $f : A \to B$ and $g : B \to A$ are both injections, then there is a set $S \subseteq A$ such that*
$$g(B - f(S)) = A - S.$$

Proof Let $h = g \circ f : A \to A$ and let $T = A - g(B)$. Now take
$$S = T \cup h(T) \cup h(h(T)) \cup \ldots$$
and note that $S = T \cup h(S)$. So since g is an injection, for any $y \in B$,
$$y \notin f(S) \Leftrightarrow g(y) \notin g \circ f(S) = h(S).$$
Since $g(y) \notin T$, this means $g(y) \notin T \cup h(S) = S$. So for any $y \in B$, $y \in B - f(S) \Leftrightarrow g(y) \in A - S$; that is,
$$g(B - f(S)) = A - S.$$
as required. □

Now for the proof of the main theorem: given S as in the lemma, note that $g^{-1}(A - S) = B - f(S)$, and define a map
$$\varphi : A \to B \quad \text{by} \quad \varphi(x) = \begin{cases} f(x) & \text{if } x \in S, \text{ and} \\ g^{-1}(x) & \text{if } x \in A - S. \end{cases}$$

The properties of S mean that φ is injective, and further

$$\varphi(A) = \varphi(S \cup (A - S)) = \varphi(S) \cup \varphi(A - S)$$

since S and $A - S$ are disjoint. So

$$\varphi(A) = \varphi(S) \cup \varphi(A - S) = f(S) \cup g^{-1}(A - S) = f(S) \cup (B - f(S)) = B.$$

Hence φ is a bijection and $A \sim B$. □

The second proof can be seen as an alternative presentation of the first; but we shall present it as succinctly as we think is reasonable. For $n \in \mathbb{N}$ define A_n by:

$$A_0 = A - g(B),$$
$$A_{n+1} = g \circ f(A_n),$$

where we are assuming that $f : A \to B$ and $g : B \to A$ are the given injections as before.

Then define $\varphi : A \to B$ by:

$$\varphi(x) = \begin{cases} f(x) & \text{if } x \in A_n \text{ for some } n; \\ g^{-1}(x) & \text{otherwise.} \end{cases}$$

φ will be the required bijection: it is defined for all $x \in A$ since g^{-1} is defined except on A_0. It will be onto B since if $y \in B$, then $g(y) \notin A_0$; so if $g(y) \in A_n$ then $n = m + 1$ and for some $x \in A_m$ we must have $y = f(x)$ (since $A_n = g \circ f(A_m)$). But then $y = \varphi(x)$; i.e. y is in the range of φ. But if $g(y) \notin A_n$ for any n, then $\varphi(g(y)) = g^{-1}(g(y)) = y$ and again y is in the range of φ.

To see that φ is 1-1, since both f and g^{-1} are 1-1, we only need to consider the case when $\varphi(x) = f(x) = g^{-1}(x') = \varphi(x')$. But if $\varphi(x)$ is defined to be $f(x)$, then we must have $x \in A_n$ for some n, and since $f(x) = g^{-1}(x')$ we have $x' = g(f(x))$, so that $x' \in A_{n+1}$ and then $\varphi(x') = f(x')$ by definition of φ. This would give $f(x) = f(x')$ and hence $x = x'$. (In fact this last case cannot happen with $x \in A_n$; but it could happen that $f(x) = g^{-1}(x)$ for some $x \in A$. In that case we would have $x \notin A_n$ for any n.) □

4.2.7 Note

Both the above proofs present problems when we try to present them in the simplest notations of set theory and prove them from our axioms; they require close examination, which we shall return to in 6.2.15(4) when we have presented the recursion theorem. The difficulty lies in making definitions such as those of the set S or the sequence A_n in the simple form $\{x \in a \mid \varphi(x)\}$

CARDINALS

(which is the only form of definition that we shall justify). In effect, the recursion theorem shows that they *can* be given in this form.

Simple uses of the Schröder–Bernstein theorem are in the exercises; two of the most important are:

Corollary 4.2.8 *(i) Any two proper intervals of $\mathbb{R}$ are equivalent;*
(ii) $\mathbb{R} \sim \mathcal{P}(\mathbb{N})$.

Proof Any proper interval I of $\mathbb{R}$ has at least two points, a and b, in it, and so contains the open interval (a, b) (where we assume $a < b$). But the methods given in examples 4.1.4(v) and (vi) easily show that $(a, b) \sim \mathbb{R}$. Hence $\mathbb{R} \preccurlyeq I$. But $I \preccurlyeq \mathbb{R}$ is immediate since $I \subseteq \mathbb{R}$. So using 4.2.6 we get $\mathbb{R} \sim I$.

To get $\mathbb{R} \sim \mathcal{P}(\mathbb{N})$, we show $[0, 1) \sim \mathcal{P}(\mathbb{N})$. Given $X \subseteq \mathbb{N}$, let $f(X) = 0.a_1 a_2 a_3 \ldots$ (decimal expansion), where

$$a_i = \begin{cases} 1 & \text{if } i \in X; \\ 0 & \text{if } i \notin X. \end{cases}$$

Then $f : \mathcal{P}(\mathbb{N}) \to [0, 1)$ is an injection. And given $x \in [0, 1)$, let

$$x = 0.b_1 b_2 b_3 \ldots$$

be its binary expansion (excluding expansions ending in repeating 1's to make them unique), and define $g(x) \subseteq \mathbb{N}$ as $\{i \in \mathbb{N} \mid b_i = 1\}$. Then $g : [0, 1) \to \mathcal{P}(\mathbb{N})$ is an injection. So again by 4.2.6 we get the result. □

4.2.9 *The continuum hypothesis*

Cantor's theorem gives us many uncountable cardinals of increasing size: $\mathbb{N}$, $\mathbb{R} \sim \mathcal{P}(\mathbb{N})$, $\mathcal{P}(\mathbb{R})$, $\mathcal{P}(\mathcal{P}(\mathbb{R}))$, ... But one unanswered question immediately arises: is $\mathbb{R}$ the next infinite cardinality above $\mathbb{N}$? Cantor raised the question, and conjectured that the answer was "yes" (he made many attempts to prove it). Eventually the conjecture that there is no cardinal strictly between $\mathbb{N}$ and $\mathbb{R}$ became known as the *continuum hypothesis*, and it was further generalized to the conjecture that for any infinite set X, there is no cardinal strictly between the cardinal of X and that of $\mathcal{P}(X)$ (the *generalized continuum hypothesis*).

It is now known that these questions are independent of the axioms we have given for set theory. Gödel [Göd38] showed, assuming that the axioms are consistent, that there could not be a proof of the negation of the generalized continuum hypothesis—in effect, he showed how to take a model of the axioms and make from it a model of the axioms together with the generalized continuum hypothesis. Then Cohen [Coh63a] showed that the continuum hypothesis could not be proved from the axioms, again assuming that the axioms are consistent. The methods developed by Gödel and Cohen have become the basis for much of modern advanced set theory, and are the subject of chapter 8.

It should be clear that the form of the results of Gödel and Cohen as stated above is the best we can hope for: they are usually referred to as *relative consistency results*, and they have to make the assumption that the axioms of set theory are consistent. For if these axioms are in fact not consistent, then everything, including both the generalized continuum hypothesis *and* its negation, will be provable, and nothing will be independent; and any proof that any result is *not* provable will prove *a fortiori* that the axioms are consistent, which by Gödel's second incompleteness theorem can only be done by assuming something at least as strong as the consistency of the axioms (see 9.6).

4.2.10   Exercises

(1) Show that for any two sets A and B, if $A \sim \mathbb{R}$ and $B \sim \mathbb{R}$ then $A \cup B \sim \mathbb{R}$.

Show further that any countable union of sets equivalent to $\mathbb{R}$ will be equivalent to $\mathbb{R}$.

[Use the Schröder–Bernstein theorem and that $\mathbb{R}^+ \sim \mathbb{R}$ and $\mathbb{R}^- \sim \mathbb{R}$. Note that there will be a naive use of the axiom of choice in the second proof, as in 4.1.12.]

(2) Show that $\mathbb{R} \times \mathbb{R} \sim \mathbb{R}$.

[One way is to show that $[0,1] \times [0,1] \preccurlyeq [0,1]$, where $[0,1]$ is the closed unit interval, and then use the Schröder–Bernstein theorem. For this, take a decimal expansion $0.a_1 a_2 a_3 \ldots$ and split it into the two decimal expansions $0.a_1 a_3 a_5 \ldots$ and $0.a_2 a_4 a_6 \ldots$. If we take no expansions ending in repeated zeros except for the expansion of zero itself, this will give a map from $[0,1]$ onto $[0,1] \times [0,1]$. Another proof is implied by the results in 4.3.7.]

(3) Construct a bijection between the open unit interval $(0,1)$ and the closed unit interval $[0,1]$.

[That they are equivalent is in 4.2.8(i); one way to get a mapping is to follow the proof implied there, using the proof of the Schröder–Bernstein theorem.]

4.3   The arithmetic of cardinal numbers

We conclude this chapter by presenting a little more of the theory of cardinals in its naive form, very much as it was first developed.

First, we shall assume that we have a notion of "the cardinal number of

CARDINALS

a set"; we shall do this naively, without saying what these cardinal numbers are, but assuming that they have certain properties. Later, in 6.7, we shall give one possible definition for cardinal numbers which can be justified from our axioms.

For finite sets this is not something new: we have always designated the cardinal number of a singleton by one, and of a pair by two, and so on (and the cardinal number of the empty set by zero). This is naive until we say what the numbers zero, one, two, ... are within our set theory (and we do this in 6.1). What is new is that we shall now assume that we can continue to talk about the cardinal number of a set even when that set is infinite.

Notation 4.3.1 The cardinal number of a set A will be denoted by $|A|$ or by Card A (another notation is $\sharp A$). We use $\kappa, \lambda, \mu, \ldots$ as variables for cardinals (i.e. cardinal numbers). The cardinal of $\mathbb{N}$ is denoted by $\aleph_0$, and the cardinal of $\mathbb{R}$ by $\mathfrak{c}$ (or sometimes by $\aleph$).

Here $\aleph$ is Aleph, the first letter of the Hebrew alphabet, introduced into mathematics by Cantor for this purpose. $\mathfrak{c}$ is for *continuum*, and is taken from the German Fraktur font; some writers used other letters from that font ($\mathfrak{m}, \mathfrak{n}, \mathfrak{p}, \ldots$) as variables for cardinals. Cantor himself introduced the notation $\overline{\overline{A}}$ for Card A, but overlines always caused problems for printers and that notation is less common today.

What we are assuming is that these cardinals not only exist, but have the following properties:

Assumption 4.3.2 *For any sets A and B,*

(i) $|A| = |B|$ *if and only if $A \sim B$;*
(ii) $|A| \leq |B|$ *if and only if $A \preccurlyeq B$, and $|A| < |B|$ if and only if $A \prec B$.*

These can be regarded as temporary assumptions; in 6.7 we shall be able to give formal definitions for this notion and in 7.2 prove these assumptions for the formal notion. What we want to do now with this notion is to introduce the operations of cardinal arithmetic:

Definition 4.3.3 (i) If A and B are disjoint sets with $|A| = \kappa$, $|B| = \lambda$, then $\kappa + \lambda = |A \cup B|$.
(ii) If A and B are any sets with $|A| = \kappa$, $|B| = \lambda$, then $\kappa \times \lambda = |A \times B|$.
(iii) If A and B are any sets with $|A| = \kappa$, $|B| = \lambda$, then $\kappa^\lambda = |{}^B A|$.

These are the operations of cardinal addition, cardinal multiplication, and cardinal exponentiation. In (ii), $A \times B$ is the cartesian product $\{(a,b) \mid a \in A \wedge b \in B\}$, and in (iii), ${}^B A$ is the set of all functions from B to A, $\{f \mid f : B \to A\}$. As usual in ordinary arithmetic, we often write $\kappa \times \lambda$ as $\kappa.\lambda$ or just $\kappa\lambda$, leaving the symbol $\times$ to denote the cartesian product. The simple properties of these operations are as one would expect from the use of the names addition, multiplication, and exponentiation; indeed, they are all immediate generalizations of possible definitions for the ordinary arithmetic operations, and clearly they extend those definitions (which originally would be applied only to finite sets).

Proposition 4.3.4 *The cardinal operations are well-defined, and cardinal addition and multiplication are commutative and associative, and multiplication distributes over addition; i.e.*

(i) *if A', B' are other sets with $A \sim A'$, $B \sim B'$, then $A \times B \sim A' \times B'$ and $^B A \sim {}^{B'} A'$; and if A' and B' are also disjoint, $A \cup B \sim A' \cup B'$.*

And for any cardinal numbers κ, λ, μ:

(ii) $\kappa + \lambda = \lambda + \kappa$, $\kappa.\lambda = \lambda.\kappa$;
(iii) $(\kappa + \lambda) + \mu = \kappa + (\lambda + \mu)$, $(\kappa.\lambda)\mu = \kappa(\lambda.\mu)$;
(iv) $\kappa(\lambda + \mu) = \kappa\lambda + \kappa\mu$;
(v) $1.\kappa = \kappa$.

The proofs of these are left as exercises. More difficult, but still left as exercises, are the standard properties of exponentiation:

Proposition 4.3.5 *For any cardinal numbers κ, λ, μ:*

(i) $\kappa^\lambda . \kappa^\mu = \kappa^{\lambda + \mu}$;
(ii) $(\kappa.\lambda)^\mu = \kappa^\mu . \lambda^\mu$;
(iii) $(\kappa^\lambda)^\mu = \kappa^{\lambda.\mu}$.

We also have the familiar monotonicity properties:

Proposition 4.3.6 *For any cardinal numbers κ, λ, μ, if $\kappa \leq \lambda$ then:*

(i) $\kappa + \mu \leq \lambda + \mu$; $\quad \kappa.\mu \leq \lambda.\mu$;
(ii) $\kappa^\mu \leq \lambda^\mu$; $\quad \mu^\kappa \leq \mu^\lambda$.

As we have said, these operations extend the familiar, finite operations. But for infinite sets, they behave rather differently, and become in some ways very simple. Thus the examples in 4.1 and 4.2 establish:

4.3.7 Examples

(i) $1 + \aleph_0 = \aleph_0$, $\quad n + \aleph_0 = \aleph_0$;
(ii) $\aleph_0 + \aleph_0 = \aleph_0$, $\quad n.\aleph_0 = \aleph_0$ (for $n \neq 0$), $\quad \aleph_0.\aleph_0 = \aleph_0$;
(iii) $\aleph_0^n = \aleph_0$ (for $n \neq 0$), $\quad \mathfrak{c} + \mathfrak{c} = \mathfrak{c}$, $\quad \aleph_0.\mathfrak{c} = \mathfrak{c}$;
(iv) $2^{\aleph_0} = \mathfrak{c}$, $\quad n^{\aleph_0} = \mathfrak{c}$ (for $n > 1$), $\quad \aleph_0^{\aleph_0} = \mathfrak{c}$, $\quad \mathfrak{c}^{\aleph_0} = \mathfrak{c}$.

The results in (iv) follow from 4.2.8 together with:

Proposition 4.3.8 *For any set A, if $|A| = \kappa$ then $|\mathcal{P}(A)| = 2^\kappa$.*

CARDINALS

Proof The mapping which takes $f \in {}^A\{0,1\}$ to the set $X_f = \{a \in A \mid f(a) = 1\}$ is a bijection from ${}^A\{0,1\}$ to $\mathcal{P}(A)$. (The function f is then called the *characteristic function* of the set X_f.) □

Now to prove the results in 4.3.7(iv) we use the following computation:

$$2 \leq n \leq \aleph_0 = \aleph_0.\aleph_0 \leq 2^{\aleph_0}; \quad \text{hence}$$

$$\mathfrak{c} = 2^{\aleph_0} \leq n^{\aleph_0} \leq \aleph_0^{\aleph_0} \leq (2^{\aleph_0})^{\aleph_0} = 2^{\aleph_0.\aleph_0} = 2^{\aleph_0} = \mathfrak{c}; \quad \text{and so}$$

$$\mathfrak{c} = \aleph_0^{\aleph_0} = (2^{\aleph_0})^{\aleph_0} = \mathfrak{c}^{\aleph_0}.$$

In fact we shall show in 6.7.5 that (under the assumption of the axiom of choice) the operations of addition and multiplication for infinite cardinals are as simple as they could be, given the monotonicity results in 4.3.6. They are simply the maximum operation: if either of κ or λ is infinite, and neither is zero, then $\kappa + \lambda = \kappa.\lambda = \max(\kappa, \lambda)$. But exponentiation is very far from being simple; we note some of the complications in chapter 8.

4.3.9 Exercises

(1) Show that the operation of cardinal addition can always be applied; i.e. for any sets A, B there are disjoint sets A', B' with $A \sim A'$ and $B \sim B'$.
   [Use $A \times \{0\}$ and $B \times \{1\}$.]
(2) Complete the proofs of 4.3.4, 4.3.5 and 4.3.6.
   [The maps required to show 4.3.4 should be straightforward. For the rules for exponentiation in 4.3.5, e.g. the last map required will need to map a function $f : M \to {}^L K$ to a function of two variables $F : M \times L \to K$. Since for $m \in M$, $f(m) : L \to K$, we can take $F(m, l) = f(m)(l)$ for $m \in M$, $l \in L$. The other cases will need similar constructions, as will 4.3.6.]
(3) For any cardinal number $\kappa \neq 0$ show that
   (i) $\kappa^1 = \kappa$;
   (ii) $1^\kappa = 1$.

5

Order relations and ordered sets

5.1 Orderings

We first give the definitions; note that we are concerned with both *partial* and *total* orderings. As in chapter 4, we are giving the development in its naive form, and later we shall examine some aspects of its formalization. Notation for binary relations (such as orderings) is very diverse; mathematicians have, historically, used only a few binary relations and have invented new notations for each as it was introduced. $\leq$ (the natural ordering by magnitude of numbers) and $\subseteq$ (the subset relation) are the best known; $\preccurlyeq$ (the relation of dominance from chapter 4) and | (divides) are also generally used. We shall use $\preccurlyeq$ as an arbitrary order relation, but when more than one is needed, we may resort to bold letters $\mathbf{r}, \mathbf{s}$, etc. We work mainly with the weak relations which include equality (like $\leq$); but we shall feel free to switch to the strict relation, excluding equality (like $<$), without much comment (see exercise 5.1.7).

Definition 5.1.1 A binary relation $\preccurlyeq$ on a set X is an *ordering* (or an *order relation*) if it is:

(i) *reflexive*, i.e. $x \preccurlyeq x$;
(ii) *antisymmetric*, i.e. $x \preccurlyeq y \land y \preccurlyeq x \Rightarrow x = y$; and
(iii) *transitive*, i.e. $x \preccurlyeq y \land y \preccurlyeq z \Rightarrow x \preccurlyeq z$ (for all $x, y, z \in X$).

An *ordered set* or *poset* is a set X which carries an ordering $\preccurlyeq$; more formally it is the structure $\langle X, \preccurlyeq \rangle$.

A *linear* (or *total*) ordering also satisfies:

(iv) for any $x, y \in X$, $x \preccurlyeq y$ or $y \preccurlyeq x$.

Here the word *poset* was coined to emphasize that the set may be partially ordered only; but we have never seen any usage which insisted that a poset may not on some occasions be linearly ordered. The word *partial*, used of an ordered set, might imply that it was not in fact a total ordering; but it is more common to emphasize this point directly, when it is needed, by saying *not total* (or *not linear*). An ordering is not total if it has *incomparable* elements, i.e. elements x and y such that neither $x \preccurlyeq y$ nor $y \preccurlyeq x$.

5.1.2 Examples

(i) The natural ordering $\leq$ on $\mathbb{N}$, $\mathbb{Z}$, $\mathbb{Q}$ or $\mathbb{R}$ is a linear ordering relation.

(ii) $\langle \mathbb{Z}^+, \preccurlyeq \rangle$ is a poset, where $m \preccurlyeq n$ if and only if m divides n. ($m \preccurlyeq n$ is more usually written $m \mid n$ in this case.)

(iii) $\langle \mathbb{Z} \times \mathbb{Z}, \preccurlyeq \rangle$ is a poset, where $(a,b) \preccurlyeq (c,d)$ if and only if $a \leq b$ and $c \leq d$.

(iv) For any non-empty set X, $\langle \mathcal{P}(X), \subseteq \rangle$ is a poset (where $\subseteq$ is the usual set-inclusion relation).

(v) Every ordering has a *dual* or *converse* ordering (given by taking the order relation in the opposite sense, so that $\leq$ becomes $\geq$, for example).

Note that examples (ii), (iii), and (iv) are *not* linear orderings (except in case (iv) if X has not more than one element). The converse of a linear ordering will still be linear.

Definition 5.1.3 If $\langle X, \preccurlyeq \rangle$ is an ordered set, and $Y \subseteq X$ is a non-empty subset of X, then $\preccurlyeq$ induces a *sub-ordering* on Y by the natural definition: $x, y \in Y \wedge x \preccurlyeq y$.

5.1.4 Examples

In the examples above, $\mathbb{N}$ is a sub-ordering of $\mathbb{Z}$, which is a sub-ordering of $\mathbb{Q}$, which is a sub-ordering of $\mathbb{R}$. (And it is conventional to use the same symbol $\leq$ for the ordering relation in these four cases.) But note that although $\mathbb{Z}^+ \subseteq \mathbb{N}$, and $m \mid n \Rightarrow m \leq n$, it is not true that $\langle \mathbb{Z}^+, \mid \rangle$ is a sub-ordering of $\langle \mathbb{N}, \leq \rangle$ (since e.g. $3 \leq 7$ but not $3 \mid 7$).

Partially ordered sets may have many totally ordered sub-orderings; we usually just say that the subset is a totally ordered subset (implying that the same ordering relation is being used). We define:

Definition 5.1.5 A totally ordered subset of a poset is called a *chain*.

5.1.6 Example

$\langle \{2,4,8\}, \mid \rangle$ is a chain in the poset $\langle \{2,4,6,8,12\}, \mid \rangle$ (where $\mid$ is "divides"). A chain may be infinite, for example $\langle \{2, 2^2, 2^3, \ldots\}, \mid \rangle$ is a chain in $\langle \mathbb{Z}^+, \mid \rangle$.

ORDER RELATIONS AND ORDERED SETS

5.1.7 Exercise

(1) Give definitions equivalent to those in 5.1.1 in terms of a *strict* ordering, i.e. an ordering such as $<$ which is anti-reflexive.

[If the strict relation is $x \prec y$ and satisfies your definitions, then the properties of 5.1.1 should hold for the weak relation $(x \prec y \vee x = y)$.]

5.2 Some properties of ordered sets

Order-preserving maps play an important role in studying posets; indeed for any interesting mathematical structures, studying the right mappings is as important as studying just the structures (or more so).

Definition 5.2.1 Let $\langle X, \preccurlyeq \rangle$ and $\langle Y, \mathsf{s} \rangle$ be posets. A function $f : X \to Y$ is *order-preserving* if, for any $a, b \in X$, $a \preccurlyeq b$ if and only if $f(a) \mathsf{\ s\ } f(b)$.

If f is also a bijection, it is called an *order-isomorphism*, and we say that $\langle X, \preccurlyeq \rangle$ and $\langle Y, \mathsf{s} \rangle$ are *order-isomorphic*, written as $\langle X, \preccurlyeq \rangle \simeq \langle Y, \mathsf{s} \rangle$ (or often just $X \simeq Y$, when the orderings can be assumed known). The words *similarity* and *similar* are also used for order-isomorphism and order-isomorphic, respectively.

5.2.2 Examples

(i) $\mathbb{Z}^+$ and $\mathbb{E}^+$ (the set of all even positive integers), both ordered by magnitude, are order-isomorphic via the bijection $x \mapsto 2x$.
(ii) $\mathbb{R} \simeq (-\frac{\pi}{2}, \frac{\pi}{2})$ via the inverse tangent function (arctan), both sets ordered by magnitude.
(iii) $\mathbb{Z}^+ \simeq \mathbb{Z}^-$, where $\mathbb{Z}^+$ is ordered by magnitude and $\mathbb{Z}^-$ by the converse (i.e. by $\geq$), under the bijection $x \mapsto -x$.

If a notion is preserved by order-isomorphisms, then it is natural to consider it as a property of an ordering. We give some of the most useful:

Definition 5.2.3 (i) Let $\langle X, \preccurlyeq \rangle$ be a poset. If there is an element $a \in X$ such that $a \preccurlyeq x$ for all $x \in X$, then a is called the *least* or *first* or *smallest* element of X.

The dual notion: if there is an element $b \in X$ such that $y \preccurlyeq b$ for every $y \in X$, then b is called the *greatest* or *last* or *largest* element of X.

By antisymmetry of the relation $\preccurlyeq$, the least (greatest) element is unique if it exists.

(ii) An element $a \in X$ is called *minimal* if $x \preccurlyeq a$ implies $x = a$, for all $x \in X$; that is, no element of X strictly precedes a.

The dual notion: an element $b \in X$ is called *maximal* if $b \preccurlyeq x$ implies $b = x$ for all $x \in X$.

5.2.4 Examples

  (i) For any poset, its least (greatest) element, if it exists, is the only minimal (maximal) element.
 (ii) Any power-set $\mathcal{P}(X)$, ordered by $\subseteq$, has $\emptyset$ as least element and X as greatest element.
(iii) The set $\mathbb{Z}$ ordered by magnitude has no least, no greatest, no minimal and no maximal elements.
(iv) The set of all non-empty proper subsets of a set X with at least two elements, ordered by $\subseteq$, has no least element and no greatest element; but every singleton $\{x\}$ for $x \in X$ is a minimal element, and every complement of a singleton $X - \{x\}$ is a maximal element.
 (v) The set of all finite subsets of $\mathbb{N}$ (ordered by $\subseteq$) has least element $\emptyset$, but no maximal elements.
(vi) The set of all infinite subsets of $\mathbb{N}$ has no minimal elements, but it has greatest element, $\mathbb{N}$ itself.
(vii) Any totally ordered set can have at most one minimal (maximal) element, which will be its least (greatest) element.
(viii) Any non-empty finite ordered set has at least one maximal and at least one minimal element.

Definition 5.2.5 Let $\langle X, \preccurlyeq \rangle$ be a poset, and B a non-empty subset of X.

 (i) An element $a \in X$ is said to be an *upper bound* of B if $y \preccurlyeq a$ for every $y \in B$. If such an element exists, B is said to be *bounded above* in X. By transitivity it follows that any $b \in X$ with $a \preccurlyeq b$ is also an upper bound of B in X.
(ii) Further, suppose that the set of all upper bounds of B in X has a least element. Then this least element is called the *least upper bound* or *supremum* of B in X, and is usually written as $\operatorname{lub} B$ or $\sup B$ (and sometimes as $\sup_X B$, if we wish to note where the supremum is being taken).
(iii) The dual notions are *lower bound* and *greatest lower bound* or *infimum*, written as $\operatorname{glb} B$ or $\inf B$.

Note that if $\sup B$ or $\inf B$ exists, it will be unique by antisymmetry, see exercise 5.2.9(1), and it may or may not belong to B; if it belongs to B, it will be the largest or smallest element of B.

5.2.6 Examples

 (i) $\mathbb{N}$, considered as a subset of $\langle \mathbb{R}, \leq \rangle$, is unbounded above, but bounded below and its infimum, 0, is in the set.
(ii) In a linear ordering, a non-empty finite set is always bounded, and contains both its supremum and its infimum (since it will always have a

ORDER RELATIONS AND ORDERED SETS

largest and a smallest element).
(iii) The set $\{1, \frac{1}{2}, \frac{1}{3}, \ldots, \frac{1}{n}, \ldots\} \subseteq \mathbb{Q}$ is bounded; its glb, 0, is not in the set; its lub, 1, is in the set.
(iv) For the half-open interval $I = [0, 1) \subseteq \mathbb{R}$, $\{x \in \mathbb{R} \mid x \leq 0\}$ is the set of all lower bounds of I, and the glb, 0, belongs to I; $\{x \in \mathbb{R} \mid x \geq 1\}$ is the set of all upper bounds, and the lub, 1, does not belong to I.
(v) Consider the ordered set $\langle \mathcal{P}(X), \subseteq \rangle$ and let $E \subseteq \mathcal{P}(X)$ be non-empty. A lower bound of E is any subset of X which is a subset of every element of E, and the greatest lower bound will be the intersection of all the elements of E, i.e. $\inf E = \bigcap E$. Dually, $\sup E = \bigcup E$.
(vi) For any poset $\langle X, \preccurlyeq \rangle$, the following are equivalent (immediately by the way we have defined sup and inf):

(a) Every non-empty subset of X which is bounded above has a supremum.
(b) Every non-empty subset of X which is bounded below has an infimum.

This last property is given a name:

Definition 5.2.7 A poset $\langle X, \preccurlyeq \rangle$ is called *order-complete* if every non-empty subset which is bounded above has a supremum in X.

(Note that there are variants on this definition, but they will not concern us here.)

5.2.8 Example

$\mathbb{R}$ is order-complete; but $\mathbb{Q}$ is not (ordered by magnitude).

We shall return to questions about maximal elements when we consider Zorn's lemma, in 7.3.

5.2.9 Exercises

(1) Show that if $B \subseteq X$ has a supremum or an infimum, then that supremum or infimum is unique.
   [Use antisymmetry.]
(2) Show that every partially ordered set is isomorphic to some collection of sets, partially ordered by $\subseteq$.
   [If $\langle X, \preccurlyeq \rangle$ is a given partially ordered set, the isomorphic collection of subsets can be taken as subsets of X, and the simplest way is to take $X_x = \{y \in X \mid y \preccurlyeq x\}$ and show that the mapping $x \mapsto X_x$ gives an order isomorphism.]

5.3 Lattices and Boolean algebras

The concept of a lattice was suggested by considering such sets as the set of all subgroups of a group. This is partially ordered by inclusion, but it has further properties as in the next definition:

Definition 5.3.1 A poset $\langle A, \preccurlyeq \rangle$ is a *lattice* if every pair of elements $\{x, y\}$ of A has a supremum, usually called the *join* of x and y and written $x \vee y$; and an infimum, usually called the *meet* of x and y and written $x \wedge y$.

If $\langle A, \preccurlyeq \rangle$ is a lattice, and $B \subseteq A$ is a non-empty subset satisfying

$$x, y \in B \Rightarrow x \wedge y, x \vee y \in B,$$

(i.e. B is *closed* under the two operations, join and meet, of A), then B is called a *sublattice* of A (and then B is itself a lattice with respect to the induced sub-ordering).

If $\langle A, \preccurlyeq \rangle$ is a lattice with the property that every non-empty subset $B \subseteq A$ has a supremum and infimum in A, then A is called a *complete* lattice. The supremum and infimum of B are then written $\bigvee B$ and $\bigwedge B$ respectively, and called the join and meet of B.

5.3.2 Examples

(i) $\mathbb{Z}$ (ordered by magnitude) is a lattice, but not a complete lattice, since the whole set $\mathbb{Z}$ has no supremum or infimum.

Similarly $\mathbb{R}$ is a lattice; in fact any totally ordered set will be a lattice, and $x \wedge y$ will be $\min\{x, y\}$, $x \vee y$ will be $\max\{x, y\}$. These can be thought of as the thinnest of all lattices.

Note that being order-complete is not sufficient to ensure that a lattice will be a complete lattice. Thus $\mathbb{R}$ is order-complete, but not a complete lattice since the whole of $\mathbb{R}$ has no supremum or infimum.

(ii) The collection of all subgroups of a group forms a lattice; so does the collection of all normal subgroups, and the normal subgroups form a sublattice of the lattice of all subgroups. But the lattice of subgroups does not (in general) form a sublattice of the lattice of all subsets of the group (all considered under the ordering $\subseteq$).

Notice what there is to prove in (ii); it has to be checked that the meet and join operations are the same for the lattices of subgroups and of normal subgroups (but not for subsets). The meet operation for subgroups is just set intersection, and is the same for normal subgroups since the intersection of two normal subgroups is again a normal subgroup. But the join operation is not in general set union; if H and H' are subgroups of a group G, then $H \vee H'$ must be the smallest subgroup of G containing both H and H'. This will be the subgroup

ORDER RELATIONS AND ORDERED SETS

generated by $H \cup H'$, which is usually more than just $H \cup H'$. (If $h \in H$ and $h' \in H'$, then e.g. $h \circ h'$ must be in $H \vee H'$, where $\circ$ is the group operation; usually $h \circ h'$ will not be in either H or H', and so not in $H \cup H'$.)

But within the lattice of normal subgroups, the join $H \vee H'$ must mean the smallest *normal* subgroup containing both H and H', so that it needs to be checked that if these are both normal, then the subgroup we got before was in fact normal. We leave that as an exercise in group theory.

(iii) The poset $\langle \mathbb{Z}, | \rangle$, is a lattice; here the meet of two numbers is the highest common factor, and the join is the least common multiple.

Similarly the set of all positive divisors of a positive integer n, ordered by division, forms a lattice.

(iv) The set $A = \{1, 2, 3, 4, 5, 6, 7, 8, 9\}$ of non-zero digits, ordered by division, i.e. $m \mid n$, is not a lattice. For $\{3, 5\}$ has no supremum in A (since $15 \notin A$).

(v) The poset $\langle \mathbb{Z} \times \mathbb{Z}, \mathbf{r} \rangle$, where $\mathbf{r}$ is defined by: $(a, b) \; \mathbf{r} \; (x, y)$ if and only if $a = x$ and $b \leq y$, is not a lattice. For, if $a \neq x$, then for any $b, y \in \mathbb{Z}$, there is no upper bound or lower bound for the pair $\{(a, b), (x, y)\}$ under the ordering $\mathbf{r}$.

(vi) For any set X, the power-set $\mathcal{P}(X)$ ordered by $\subseteq$, is a complete lattice. Here $\wedge$ is $\cap$, set intersection, and $\vee$ is $\cup$, set union.

(vii) The set of all discs in $\mathbb{R}^2$ ordered by $\subseteq$ is not a lattice. (Although there is a good notion of the largest disc contained in the intersection of two discs, and of the smallest disc containing two discs, these will not satisfy the definitions of infimum and supremum.)

But if we restrict to discs with centres on the x-axis, we do get a lattice.

Lattice theory is important in algebra, but we are more interested in developing set theory, and for us lattice theory is an example of its application. In this context the power-set lattice is important, and it has further properties, which go to make it a Boolean algebra. The first is distributivity:

Definition 5.3.3 A lattice $\langle A, \preccurlyeq \rangle$ is *distributive* if, for all $a, b, c \in A$,

(i) $a \wedge (b \vee c) = (a \wedge b) \vee (a \wedge c)$, and
(ii) $a \vee (b \wedge c) = (a \vee b) \wedge (a \vee c)$.

In fact we only need one of these two properties, since each implies the other. We prove one of these implications as an example of the methods in lattice theory:

Theorem 5.3.4 *If $\langle A, \preccurlyeq \rangle$ is a lattice, and for all $a, b, c \in A$,*

$$a \wedge (b \vee c) = (a \wedge b) \vee (a \wedge c),$$

then also
$$a \vee (b \wedge c) = (a \vee b) \wedge (a \vee c).$$

Proof First note that in any lattice, $(a \wedge b) \preccurlyeq a$ and $a \preccurlyeq (a \vee b)$, and hence we have the trivial identity $a \vee (a \wedge b) = a$ and its dual, $a \wedge (a \vee b) = a$ (these immediately from the definition of meet and join). Also meet and join are commutative and associative. So we have:

$$\begin{aligned}
a \wedge (b \vee c) &= [a \vee (a \wedge c)] \vee (b \wedge c) \quad \text{by the trivial identity} \\
&= a \vee [(a \wedge c) \vee (b \wedge c)] \quad \text{by associativity} \\
&= a \vee [(a \vee b) \wedge c] \quad \text{by our assumption} \\
&= [(a \vee b) \wedge a] \vee [(a \vee b) \wedge c] \quad \text{by the dual trivial identity} \\
&= (a \vee b) \wedge (a \vee c) \quad \text{again by our assumption.}
\end{aligned}$$

□

5.3.5 Examples

(i) Any chain is a distributive lattice; so is any lattice of integers under division.

(ii) Any power-set is a distributive lattice; indeed, any lattice of sets, for which meet is set intersection and join is set union, will be distributive.

(iii) But the lattice of subgroups of a group is not in general distributive. (Note that here the join will not be set union, as noted in 5.3.2(ii) above.)

(iv) $\{\mathbb{O}, a, b, c, \mathbb{I}\}$ ordered as in the diagram, is not distributive (since $a \vee b = a \vee c = \mathbb{I}$, so $(a \vee b) \wedge (a \vee c) = \mathbb{I}$; but $a \wedge (b \vee c) = a \wedge \mathbb{I} = a$).

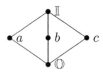

Diagram 5.3.5(iv)

The further properties of any power-set involve *complements* of elements, and structures with these properties are called *Boolean algebras* after George Boole (1815–1869).

Definition 5.3.6 A *Boolean algebra* is a distributive lattice A with the further properties:

(i) A has a least element (or *zero*, usually denoted by $\mathbb{O}$) and a greatest element (or *unit*, usually denoted by $\mathbb{I}$);

ORDER RELATIONS AND ORDERED SETS

(ii) For each element $x \in A$ there is an element $x' \in A$ such that $x \wedge x' = \mathbb{O}$ and $x \vee x' = \mathbb{I}$. This x' can be shown to be unique, see exercise 5.3.16(1), and is called the *complement* of x in A.

It is also possible to give a purely algebraic definition of a Boolean algebra. An algebraic definition uses only the algebraic operations $\wedge$, $\vee$, and $'$, and the constants $\mathbb{O}$ and $\mathbb{I}$ (and defines the order relation from them, by a definition such as $x \preccurlyeq y$ if and only if $x \vee y = y$), in contrast to the definition above which uses only the order relation $\preccurlyeq$ as primitive and defines the constants $\mathbb{O}$ and $\mathbb{I}$ and the three operations $\wedge$, $\vee$, and $'$, in terms of $\preccurlyeq$. See exercise 5.3.16(2).

5.3.7   Examples

Diagram 5.3.7(i)

(i) $\{\mathbb{O}, b, c, \mathbb{I}\}$, ordered as in the diagram, is a Boolean algebra (with $b' = c$ and $c' = b$).
(ii) For any set X, the poset $\langle \mathcal{P}(X), \subseteq \rangle$ is a Boolean algebra, with $\emptyset$ as $\mathbb{O}$, X as $\mathbb{I}$, and $x' = X - x$.
(iii) The set of positive divisors of a positive integer n, under division, is a Boolean algebra, with 1 as $\mathbb{O}$ and n as $\mathbb{I}$, provided n is *square-free*, i.e. provided no prime p has $p^2 \mid n$. (Otherwise complements cannot be defined.)

The power-set Boolean algebras have yet one more property, namely they are complete as lattices (as noted above). For finite Boolean algebras this is not a restriction; every finite lattice is complete as a lattice, and every finite Boolean algebra is isomorphic to a power-set, see exercise 5.3.16(3). But for infinite Boolean algebras, this is a real restriction; there are many infinite Boolean algebras which are not isomorphic to any power-set. This can be seen easily when we remember that no power-set can be countably infinite (see 4.2.3).

5.3.8   Example

The collection of all subsets A of a non-empty set X such that either A is finite or $X - A$ is finite, forms a Boolean algebra (ordered by $\subseteq$). (A set A for which is $X - A$ is finite is called *cofinite in X*, or just *cofinite* if X is clear.) If X is infinite, this Boolean algebra is not complete, and if X is countable, e.g. $X = \mathbb{N}$, this will be a countably infinite Boolean algebra.

Not every Boolean algebra is isomorphic to a power-set; but the example above does hint at what can be proved about every Boolean algebra. We showed in exercise 5.2.9(2) that every poset is similar to some collection of sets, ordered by $\subseteq$; and we can extend that result when we consider Boolean algebras so that not only the ordering but also the operations are the familiar set-theoretic operations (i.e. $\cap$, $\cup$ and set complement). We use:

Definition 5.3.9 A non-empty collection A of subsets of a set X is called a *set algebra* if A is closed under union, intersection, and complementation. It is a *set lattice* if it is just closed under union and intersection.

Note that a set algebra is automatically a Boolean algebra under the relation $\subseteq$.

Now we can state:

Theorem 5.3.10 (The Stone representation theorem) *Every Boolean algebra is isomorphic to some set algebra.*

We shall outline the ideas needed to prove this, since they introduce some important notions, and also lead to an application of the axiom of choice which will be presented in 7.4.5. First the notions of ideals and filters:

Definition 5.3.11 Let $\langle A, \preccurlyeq \rangle$ be a lattice. A non-empty subset $I \subseteq A$ is an *ideal* of A if, for any $a, b \in A$,

(i) $a \in I$ and $b \in I$ together imply $a \vee b \in I$; and
(ii) if $b \in I$ and $a \preccurlyeq b$ then $a \in I$.

It can help to think of the members of an ideal I as "small elements" of the lattice, in a new sense of "small", which must agree with the ordering on the lattice. Thus if a and b are both "small", so is $a \vee b$; and if b is "small", so is any $a \preccurlyeq b$.

5.3.12   Example

Let X be $\mathcal{P}(\mathbb{N})$ and let I be the set of all finite subsets of $\mathbb{N}$.

The dual notion is:

Definition 5.3.13 Let $\langle A, \preccurlyeq \rangle$ be a lattice. A non-empty subset $F \subseteq A$ is a *filter* in A if, for any $a, b \in A$,

(i) $a \in F$ and $b \in F$ together imply $a \wedge b \in F$; and
(ii) if $a \in F$ and $a \preccurlyeq b$ then $b \in F$.

Just as elements of an ideal can be thought of as "small", elements of a filter can be thought of as "large"; and the most typical example of a filter would be the filter of cofinite subsets of $\mathbb{N}$. Filters and ideals are called *proper* if they are not the whole lattice, and this is so commonly required that it is sometimes included in the definition. But then we run into awkwardness when we want to define ideals or filters generated by subsets, as in 5.3.16(5) below.

ORDER RELATIONS AND ORDERED SETS   61

5.3.14   Examples

(i) If $\langle A, \preccurlyeq \rangle$ is a lattice and $a \in A$ then the set $\{x \in A \mid x \preccurlyeq a\}$ is an ideal of A, called the *principal ideal* generated by a. It will be proper unless a is the greatest element of A. Dually, the set $\{x \in A \mid a \preccurlyeq x\}$ is a filter, the *principal filter* generated by a, which will be proper unless a is the least element in A.

(ii) In the lattice $\langle \mathbb{Z}^+, | \rangle$ of 5.3.2(iii), for any $x \in \mathbb{Z}^+$, the set $x\mathbb{Z}^+ = \{xz \mid z \in \mathbb{Z}^+\}$ is a filter, proper if $x \neq 1$. Further, if x is prime, then the filter $x\mathbb{Z}^+$ is a *prime filter*, where a filter F in a lattice A is *prime* if F is proper and, whenever $a \vee b \in F$ for $a, b \in A$, then either $a \in F$ or $b \in F$.

Note how this last example points out a confusion with the usual definitions given in number theory: the set $x\mathbb{Z}^+ = \{xz \mid z \in \mathbb{Z}^+\}$ is always called an *ideal* by number theorists, but we have called it a filter. How can this confusion arise? Very simply: we have taken the ordering converse to the ordering which the number theorist is thinking of. In fact there are good reasons within number theory to think of division as giving an ordering which is the other way round to the ordering by magnitude; examples of this occur when considering p-adic numbers. But since we are looking at these orderings only as examples of set theory, we have kept to the simple-minded ordering for division which agrees with magnitude.

We can now sketch the following theorem, which implies 5.3.10:

Theorem 5.3.15 *Every distributive lattice is isomorphic to a lattice of sets.*

Proof Let $\langle A, \preccurlyeq \rangle$ be the given distributive lattice; we map each element $a \in A$ to the set $\mathcal{I}_a$ of all prime ideals of A which do not contain a (where an ideal I is a *prime ideal* if I is proper and, whenever $a \wedge b \in I$, then either $a \in I$ or $b \in I$; in other words the dual of a prime filter). Then we show that $\{\mathcal{I}_a \mid a \in A\}$ is a lattice of sets, and that this mapping is an isomorphism. The main steps, most of which we leave as exercises, are:

(i) To show the mapping is 1-1. If $a, b \in A$ with $a \neq b$ then we must show $\mathcal{I}_a \neq \mathcal{I}_b$; in particular, if $a \npreccurlyeq b$ with $a \neq b$, there is a prime ideal I in A with $a \in I$ and $b \notin I$ (so that $I \in \mathcal{I}_b - \mathcal{I}_a$). Distributivity of A is needed here if the ideal I is to be prime. The proof of this step requires the axiom of choice, and is given in 7.4.5.

(ii) Hence the mapping is an order-isomorphism, since $a \preccurlyeq b$ implies $\mathcal{I}_a \subseteq \mathcal{I}_b$ immediately from the definition of $\mathcal{I}_a$.

(iii) We now show that $\mathcal{I}_{a \vee b} = \mathcal{I}_a \cup \mathcal{I}_b$ and $\mathcal{I}_{a \wedge b} = \mathcal{I}_a \cap \mathcal{I}_b$. Here it is essential that we use prime ideals; to show that $\mathcal{I}_a \cap \mathcal{I}_b \subseteq \mathcal{I}_{a \wedge b}$, suppose that $I \in \mathcal{I}_a \cap \mathcal{I}_b$. Then $a \notin I$ and $b \notin I$; so, assuming I is prime, we get $a \wedge b \notin I$ and so $I \in \mathcal{I}_{a \wedge b}$ as required.

(The other three inclusions needed all follow from the definition of an ideal, and are left as exercises.) □

Some further developments of the theory of ideals are indicated in the exercises.

5.3.16 Exercises

(1) Show that complements in a Boolean algebra are unique. Show further that in any distributive lattice, relative complements are also unique, where if a, b, x and y are in a distributive lattice, with $a \preccurlyeq x \preccurlyeq b$ and $a \preccurlyeq y \preccurlyeq b$, then x and y are *complements relative to a and b* if $x \wedge y = a$ and $x \vee y = b$.

Show that this can fail if the lattice is not distributive.

Show that Boolean algebras always have relative complements.

[Suppose that z, with $a \preccurlyeq z \preccurlyeq b$ is another relative complement of x, i.e. that also $x \wedge z = a$ and $x \vee z = b$. Then use distributivity on $y = y \wedge b = y \wedge (x \vee z)$ to get $y = z$. The example in 5.3.5(iv) is a typical non-distributive lattice and can be used to show the failure of uniqueness of complements, and most other failures in non-distributive lattices. But note also the work on non-modular lattices, which are also non-distributive, in exercise (11) below.]

(2) Algebraic definition of lattices and Boolean algebras. Show that the following gives an equivalent definition of a lattice:

A lattice is a structure $\langle A, \vee, \wedge \rangle$ where $\vee$ and $\wedge$ are binary operations which are both commutative and associative and further satisfy:

(i) idempotency: $a \vee a = a$ and $a \wedge a = a$;
(ii) and the identities: $a \wedge (a \vee b) = a$ and $a \vee (a \wedge b) = a$.

[Show that if the definition of order is taken to be: $a \preccurlyeq b$ if and only if $a \wedge b = a$, then transitivity follows immediately from the associativity of $\wedge$, reflexivity follows from its idempotency, and antisymmetry follows from its commutativity. Further the fact that $a \wedge b$ is the infimum of the pair a, b for the ordering $\preccurlyeq$ follows from all of these properties of $\wedge$.

Then show that the fact that $a \vee b$ is the supremum of the pair a, b for the ordering $\preccurlyeq$ follows from the two identities in (ii). Show first that these identities imply that $a \preccurlyeq b$ if and only if $a \vee b = b$.]

Now show that the following gives an equivalent definition of a Boolean algebra:

A Boolean algebra is a structure $\langle A, \vee, \wedge, ', \mathbb{O}, \mathbb{I} \rangle$ such that $\langle A, \vee, \wedge \rangle$ forms a lattice, and $'$ is a unary operation and $\mathbb{O}$ and $\mathbb{I}$ are constants such that

(iii) $a \vee \mathbb{O} = a$ and $a \wedge \mathbb{O} = \mathbb{O}$, and $a \vee \mathbb{I} = \mathbb{I}$ and $a \wedge \mathbb{I} = a$; and
(iv) $a \vee a' = \mathbb{I}$ and $a \wedge a' = \mathbb{O}$.

ORDER RELATIONS AND ORDERED SETS 63

(3) *Atoms* in a Boolean algebra. If $\langle A, \preccurlyeq \rangle$ is a Boolean algebra and $a \in A$ is a non-zero element with the property

$$x \preccurlyeq a \Rightarrow x = \mathbb{O} \text{ or } x = a$$

then a is called an *atom* of A.

Show that in every finite Boolean algebra, each non-zero element b is the join of all atoms $\preccurlyeq b$. Hence show that any finite Boolean algebra is isomorphic to a power-set (namely the power-set of the set of all atoms).

Show that these properties can fail in infinite Boolean algebras.

[The simplest Boolean algebra which has no atoms at all is the Boolean algebra which is the set algebra generated by half-open intervals (say, open on the left and closed on the right) in $\mathbb{R}$ or in the half-open unit interval $(0, 1]$. If we take only intervals with rational endpoints, we get a countable atomless Boolean algebra.]

(4) Complete the proof of theorem 5.3.15 part (iii) by showing that $\mathcal{I}_{a \wedge b} \subseteq \mathcal{I}_a \cap \mathcal{I}_b$, $\mathcal{I}_{a \vee b} \subseteq \mathcal{I}_a \cup \mathcal{I}_b$ and $\mathcal{I}_a \cup \mathcal{I}_b \subseteq \mathcal{I}_{a \vee b}$.

[As noted, these all follow from the definition of an ideal.]

Show further that if a lattice B is not a linear ordering, then the mapping from $b \in B$ to the set $\mathcal{J}_b$ of all ideals not containing b, is not an isomorphism. (So that it was essential to use *prime* ideals, and not just *ideals*, for the theorem.)

[The property which fails is that we shall not get $\mathcal{J}_{a \wedge b} = \mathcal{J}_a \cap \mathcal{J}_b$; the principal ideal generated by $a \wedge b$, which is not in $\mathcal{J}_{a \wedge b}$, will be in $\mathcal{J}_a \cap \mathcal{J}_b$ if a and b are incomparable in B.]

(5) Extensions of ideals. Given an ideal I and an element $a \notin I$, the smallest ideal which contains both I and a is called the *extension* of I by a (sometimes denoted by $I + a$). Show that an element b is in this extension if and only if there is an element $c \in I$ such that $b \preccurlyeq a \vee c$.

More generally, given any subset X of a lattice A, the *ideal generated by X* is the smallest ideal in A which contains X. Show that an element b is in this ideal generated by X if and only if there is a finite subset $\{x_1, \ldots, x_n\}$ of X such that $b \preccurlyeq x_1 \vee \ldots \vee x_n$.

What will be the dual notions for filters?

[Note that if these notions are to be completely general, we have to allow the improper ideal or filter, since the extension of an ideal or the ideal generated by a set can easily be improper.]

(6) Prime ideals and maximal ideals. A *maximal* ideal in a lattice is a proper ideal I such that no proper ideal properly contains I (hence maximal in the collection of all proper ideals, ordered by inclusion). A *prime* ideal is defined in 5.3.15, and contains either a or b whenever it contains $a \wedge b$.

Show that in a Boolean algebra, these notions coincide. Hence show that 5.3.10 follows from 5.3.15.

However, show that in a general lattice they are different notions. If A is a distributive lattice, show that a maximal ideal must be prime, but a prime ideal may not be maximal. And show that in a non-distributive lattice, a maximal ideal may not be prime.

[To show that in a distributive lattice, an ideal which is not prime cannot be maximal, suppose that $a \wedge b \in I$ but $a \notin I$, $b \notin I$ and show that the extension of I by a will not contain b, and vice versa (since if $b \trianglelefteq a \vee c$ for some $c \in I$, then $(a \wedge b) \vee c \in I$; and $b \trianglelefteq (a \vee c) \wedge (b \vee c) = (a \wedge b) \vee c$ using distributivity). For an example of an ideal which is prime but not maximal, one can use a linear ordering: show that every ideal in a lattice which is a linear ordering will count as prime.

Now to show that the notions coincide in a Boolean algebra, use the fact that in a Boolean algebra, $\mathbb{O}$ is in every ideal, and $\mathbb{O} = a \wedge a'$ for every $a \in A$. And show that an ideal is maximal if and only if for each element a of the Boolean algebra, either a or a' (and not both) is in the ideal.

The example of 5.3.5(iv) is a non-distributive lattice which has e.g. $\{\mathbb{O}, a\}$ as a maximal ideal which is not prime.]

(7) Show that if every ideal in a lattice with $\mathbb{O}$ is principal, then the lattice is a complete lattice.

[An example of such a lattice which is not finite would be the collection of cofinite subsets of an infinite set, together with $\emptyset$, under inclusion.]

But show that a complete lattice can still have non-principal ideals.

[In the power-set Boolean algebra of an infinite set, the ideal of finite subsets will not be principal.]

(8) Let A be the collection of all finite open intervals in $\mathbb{Q}$, together with the empty set, ordered by $\subseteq$. Show that this is a lattice with no prime ideals and no maximal ideals.

[Note that meet will be intersection, but the join of two elements (intervals) of A will not usually be union but the smallest open interval containing both the intervals. Show that any proper ideal I must be such that for some $\alpha \in \mathbb{Q}$, every interval in I is entirely to the left of α (or all are entirely to the right). $\emptyset$ will be in any ideal, so if an ideal were prime it would contain at least one of any pair of disjoint intervals.]

(9) Regular open algebras. Given any partial ordering $\langle P, \trianglelefteq \rangle$, define a topology on P by: a set $A \subseteq P$ is open if $x \in A \wedge x \trianglelefteq y \Rightarrow y \in A$. An alternative description is: for $x \in P$ let $x^{\trianglelefteq} = \{y \in P \mid x \trianglelefteq y\}$ (the *cone* above x in P) and say A is open if $x \in A \Rightarrow x^{\trianglelefteq} \subseteq A$ (so the cones form a basis for the topology). Then say that $A \subseteq P$ is *regular open* (RO) if A is equal to the interior of its closure in this topology.

Show that A is RO if A is open and $x^{\trianglelefteq} \subseteq \operatorname{cl} A \Rightarrow x \in A$ for $x \in P$ (where $\operatorname{cl} A$ is the closure of A; show that $\operatorname{cl} A = \{z \in P \mid \exists y \in A(z \trianglelefteq y)\}$).

ORDER RELATIONS AND ORDERED SETS

Now show that the regular open subsets of P form a complete Boolean algebra under inclusion.

[The meet operation will be intersection, but the join operation will be given by int cl$(A \cup B)$, or by int cl$\bigcup_{i \in I} A_i$ for a family $(A_i)_{i \in I}$ of RO sets. Show that int$(P \setminus A)$ will act as complement.

This is an important source of complete Boolean algebras; its main use is in forcing and independence proofs. Some further notes are in 8.14. But topology is not needed for the rest of this book.]

(10) Some product lattices. Show that $\mathbb{R} \times \mathbb{R}$ ordered by $(a,b) \preccurlyeq (x,y) \Leftrightarrow a \leq x \wedge b \leq y$ is a lattice, and is distributive.

Similarly for every set A, show that the set $F = {}^A\mathbb{R}$, ordered by $f \leq g \Leftrightarrow \forall x \in A(f(x) \leq g(x))$, is a distributive lattice.

For the case A is the closed unit interval $[0,1]$, let E be the subset of all continuous functions in F. Show that E is a sublattice of F.

(11) Modular lattices. A lattice $\langle A, \preccurlyeq \rangle$ is called *modular* if

$$a \preccurlyeq b \Rightarrow a \vee (b \wedge c) = b \wedge (a \vee c) \quad \text{for any } a,b,c \in A.$$

Show that $a \preccurlyeq b \Rightarrow a \vee (b \wedge c) \preccurlyeq b \wedge (a \vee c)$ in any lattice; and that a distributive lattice must be modular. Show that the lattice of 5.3.5(iv) is modular but not distributive. And show that a lattice is not modular if and only if the lattice in the diagram below can be embedded in it (preserving meets and joins, and with $a \neq b$).

[Most of the lattices considered so far are modular. Exceptions are the lattice in (8) above, and the lattice of subgroups of a group; abelian groups have modular lattices of subgroups, but e.g. the symmetric group S_4, and also the infinite dihedral group, do not have modular lattices of subgroups.]

Show that a lattice is not *distributive* if and only if either the five-element lattice below, or the lattice of 5.3.5(iv), can be embedded in it.

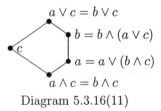

Diagram 5.3.16(11)

Further exercises on ideals and filters require the axiom of choice and are in 7.4.6.

5.4 Well-ordered sets

Definition 5.4.1 A poset $\langle X, \preccurlyeq \rangle$ is *well-ordered* if and only if every non-empty subset of X has a least element.

Note immediately that every well-ordered set must be totally ordered (just consider all pairs of elements; they must each have a least member). Also the whole set X must have a least member; and every subset of X will also be well-ordered by the induced sub-ordering.

5.4.2 Examples

(i) $\mathbb{N}$ and $\mathbb{Z}^+$, ordered by magnitude, are well-ordered.
(ii) Any finite linearly ordered set is well-ordered.
(iii) $\mathbb{N}^*$, which is $\mathbb{N}$ ordered by the converse of the usual ordering by magnitude, is not well-ordered. This is the typical non-well-ordered set; $\mathbb{Z}^-$, the negative integers ordered by magnitude, is isomorphic. Every linearly ordered, non-well-ordered set must contain such a *descending chain*; though, as we shall see, we need the axiom of choice to prove this (see 7.4.4).
(iv) $\mathbb{Z}$, $\mathbb{Q}$, $\mathbb{R}$, or any interval in $\mathbb{Q}$ or $\mathbb{R}$, ordered by magnitude, is not well-ordered (we can easily define descending sequences such as $(\frac{1}{n})$ in $\mathbb{Q}$).
(v) $\mathbb{Z}^+$, written as $\{1, 3, 5, \ldots, 2, 4, 6, \ldots\}$, is a well-ordering, i.e. when ordered by $\preccurlyeq$ defined by $x \preccurlyeq y$ if and only if (a) either x is odd and y is even; or (b) x and y have the same parity and $x \leq y$.

Note that this well-ordering of $\mathbb{Z}^+$ is not isomorphic to the usual well-ordering of $\mathbb{Z}^+$; this is seen by noting that in the usual ordering of $\mathbb{Z}^+$, there is just one element, 1, which has no immediate predecessor; whereas in the new ordering $\preccurlyeq$ just defined there are two (1 and 2).

Cantor gave an informal development of the theory of well-orderings, which was parallel (and prior) to his theory of cardinals: just as he used one–one mappings to define (informally) the notion of a cardinal for any set, so he used order-isomorphisms to define (informally) the notion of *ordinal* for any well-ordered set. If we consider the distinction made in natural language between cardinal numbers, in English *one, two, three*, ..., (used for counting things), and ordinal numbers, in English *first, second, third*, ..., (used for ordering things), it is clear that we are dealing with two different, but related ideas. Cantor showed that both notions can be extended into the transfinite, and gave both developments in the same style, somewhat as we have set out for the cardinals in chapter 4.

We shall not present here the development of ordinals in this informal style, since we are more concerned to show how the development can be given within ZF set theory, which is presented in chapter 6, and we do not wish to duplicate

ORDER RELATIONS AND ORDERED SETS

so much of the work. (Some of the informal development is indicated in the exercises 5.4.6.)

We present just one important result here which illustrates well how the definition of well-ordering can be used, and which also shows how different well-orderings are from general linear orderings.

Theorem 5.4.3 *Any order-preserving map from a well-ordered set to itself must be increasing; i.e. if $\langle X, \preccurlyeq \rangle$ is a well-ordering, and $f : X \to X$ is order-preserving, then $x \preccurlyeq f(x)$ for all $x \in X$.*

Proof This is a proof *by induction on the well-ordering*, and this notion of *induction* is the powerful new method which the notion of well-ordering makes available. We set it out here in one way, but in chapter 6 we shall see other ways to set out arguments by induction which will be closer to the usual way that *mathematical induction* is introduced. The way we are setting out here is sometimes called the *least number principle*.

Suppose that the theorem fails; i.e. suppose that we have a well-ordering $\langle X, \preccurlyeq \rangle$ and a mapping $f : X \to X$ which is not increasing. We shall show that f cannot be order-preserving.

So we take the set $Y = \{x \in X \mid x \npreceq f(x)\}$, and since Y is a non-empty (by assumption) subset of X, and $\preccurlyeq$ well-orders X, Y must have a least member, say y, under $\preccurlyeq$. Now let $z = f(y)$ and consider the action of f on y and z. $y \in Y$, so $y \npreceq f(y)$, i.e. $y \npreceq z$ and so we must have $z \preccurlyeq y$ and $z \neq y$. Hence z cannot be in Y, since y is $\preccurlyeq$-least in Y. But $z \in X$, so we must have $z \preccurlyeq f(z)$, and this shows that f is not order-preserving (we have found $y, z \in X$ with $z \neq y$, and $z \preccurlyeq y$ but $f(y) \npreceq f(z)$, since $z = f(y)$). □

Corollary 5.4.4 *No well-ordering is isomorphic to a proper initial segment of itself. Hence distinct initial segments of a well-ordering are not isomorphic.*

Proof Here for any ordering $\langle X, \preccurlyeq \rangle$, an *initial segment* is a subset which is "closed downwards" in the ordering; i.e. a subset $Y \subset X$ such that if $y \in Y$, and $x \in X$ with $x \preccurlyeq y$, then $x \in Y$.

So suppose that Y is a proper initial segment of the well-ordering $\langle X, \preccurlyeq \rangle$, and $f : X \to Y$. Then if $x \in X - Y$, we must have $f(x) \preccurlyeq x$ and so by 5.4.3, f cannot be an order-preserving map (and hence not an order-isomorphism). □

5.4.5 Example

Note that there can be many order-preserving maps of well-orderings which *are* increasing; e.g. the map $f : \mathbb{N} \to \mathbb{N}$ given by $f(n) = 2n$. And for non-well-orderings, a proper initial segment *can* be isomorphic to the whole; e.g. $\mathbb{R}^-$, the set of negative reals, is order-isomorphic to $\mathbb{R}$.

5.4.6 Exercises

(1) Addition for well-ordered sets. Show that if $\langle X, \preccurlyeq \rangle$ and $\langle Y, \mathsf{s} \rangle$ are disjoint and both well-ordered, then $\langle X \cup Y, \mathsf{t} \rangle$ is also a well-ordering, where $a \,\mathsf{t}\, b$ holds just when:

$$a \in X \text{ and } b \in X \text{ and } a \preccurlyeq b; \text{ or}$$
$$a \in X \text{ and } b \in Y; \text{ or}$$
$$a \in Y \text{ and } b \in Y \text{ and } a \,\mathsf{s}\, b.$$

Write $X + Y$ for $\langle X \cup Y, \mathsf{t} \rangle$, and show that $n + \mathbb{N} \simeq \mathbb{N}$ but $\mathbb{N} + n \not\simeq \mathbb{N}$ for $n \geq 1$.

[Here, when we want to think of n as a well-ordering, we take it to be the well-ordering of $\{0, 1, 2, \ldots, n-1\}$ under the natural order, or anything order-isomorphic to this. (As in 4.3.9(1) we can always find isomorphic well-orderings which are disjoint.) To show that an ordering is isomorphic to $\mathbb{N}$, note that we only need to show that it is infinite and has the property that *every element has only finitely many predecessors* under the ordering.]

Show that $\mathbb{N} + \mathbb{N}$ is order-isomorphic to the ordering in example 5.4.2(v).

[This is the *sum* of the two well-orderings.]

(2) Multiplication for well-ordered sets. Show that if $\langle X, \preccurlyeq \rangle$ and $\langle Y, \mathsf{s} \rangle$ are both well-ordered, then $\langle X \times Y, \mathsf{t} \rangle$ is also a well-ordering, where $\langle x, y \rangle \,\mathsf{t}\, \langle x', y' \rangle$ holds just when:

$$y \,\mathsf{s}\, y'; \text{ or } y = y' \text{ and } x \preccurlyeq x'.$$

Write $X \times Y$ for $\langle X \times Y, \mathsf{t} \rangle$, and show that $n \times \mathbb{N} \simeq \mathbb{N}$ for $n \geq 1$, but that $\mathbb{N} \times 2 \simeq \mathbb{N} + \mathbb{N}$.

[This is the *product* of the two well-orderings. This ordering of the product is sometimes called *ordering by last difference* or the *reverse lexicographic* ordering, since it follows the rules used for ordering words in a dictionary, except that they are applied from right to left instead of from left to right (the dictionary would use *first* difference).]

(3) Continued products for families of well-orderings. Let $\langle X_i, \preccurlyeq_i \rangle$ be well-orderings, for $i \in I$, and let the index set I also be well-ordered, by s. Let $\prod' X_i$ be the *restricted product* of this family, i.e. the set

$$\{f : I \to \bigcup X_i \mid \forall i \in I (f(i) \in X_i) \wedge \{i \in I \mid f(i) \neq 0_i\} \text{ is finite}\}$$

ORDER RELATIONS AND ORDERED SETS   69

where 0_i is the least member of X_i under $\preccurlyeq_i$ for each $i \in I$. (The set $\{i \in I \mid f(i) \neq 0_i\}$ is called the *support* of the function f, and we restrict the product to functions with *finite support*.)

Now let **t** be the ordering of $\prod' X_i$ given by the reverse lexicographic order: if $f, g \in \prod' X_i$, then $f \mathbf{t} g$ if and only if $f(i_1) \preccurlyeq_{i_1} g(i_1)$ for that $i_1 \in I$ which is the *last difference* of f and g, i.e. for which $f(i_1) \neq g(i_1)$ but for every $i' \in I$ with $i_1 \mathbf{s} i'$, we have $f(i') = g(i')$. (Note that is essential to confine ourselves to the restricted product, with only finite supports, in order that this last difference should always exist.)

Show that the restricted product $\prod' X_i$ is well-ordered by **t**.

[This can be done directly, or by using the axiom of choice and showing that there cannot be a descending chain. Working directly, show that if $Y \subseteq \prod' X_i$ is not empty, then Y has a least member: successively restrict attention to smaller and smaller subsets of Y, until we have found the least member. First, for $f \in Y$ consider the greatest member of the support of f, and restrict to those $f \in Y$ for which this is least, say i_0, to get $Y_0 \subseteq Y$. Next for $f \in Y_0$, consider the values of f at i_0, and restrict to those f for which this is least (say x_0) to get $Y_0' \subseteq Y_0$. Next for $f \in Y_0'$ consider the second greatest member of its support, and restrict to those f for which this is least, say i_1, to get $Y_1 \subseteq Y_0'$, then restrict to Y_1' by considering the values at i_1, and so on. Stop when the restricted set so obtained contains just one member.

Show that this process must in fact stop after finitely many steps, and that the single member thus obtained will be the least member of Y under **t**. (Use induction to show that at every stage, if $f \in Y$ is retained while $g \in Y$ is removed, then $f \mathbf{t} g$.)

To show directly that there cannot be a descending chain in the ordering **t**, show that an infinitely descending chain $\ldots \mathbf{t} f_n \mathbf{t} f_{n-1} \mathbf{t} \ldots \mathbf{t} f_1 \mathbf{t} f_0$ in **t** implies an infinitely descending chain either in **s** or in $\preccurlyeq_i$ for some $i \in I$. Steps of restriction similar to those above will be needed.]

Show that $\prod' n$ taken over $\mathbb{N}$ as index is isomorphic to $\mathbb{N}$.

(4) Show that the *direct* lexicographic ordering of the product, even the restricted product, of infinitely many well-orderings, indexed by a well-ordering, will not in general give a well-ordering.

[Take all the factors as the ordering 2 and the index set as $\mathbb{N}$, and show that we can represent the binary decimal expansions of the descending sequence $\frac{1}{2}, \frac{1}{4}, \frac{1}{8}, \ldots$ within this product, each with a support of size one.]

(5) Exponentiation for well-ordered sets. This can now be given directly from the definition of continued products; to get the well-ordering which represents $\langle X, \preccurlyeq \rangle$ raised to the power $\langle Y, \mathbf{s} \rangle$ (where these are both well-orderings), we take the continued product of the constant family with value X, indexed by Y.

Write this as X^Y; but note that it is a *different* operation to the cardinal exponentiation of 4.3. Show that if X and Y are finite, this gives the usual notion of exponentiation for natural numbers. But show that $n^{\mathbb{N}} \simeq \mathbb{N}$ for any $n \geq 2$, and $\mathbb{N}^2 \simeq \mathbb{N} \times \mathbb{N}$, for this form of exponentiation.

These exercises give the informal notions of *ordinal arithmetic*, and in 6.6.8 they will be given formal counterparts. The notions make sense for arbitrary linear orderings and not just well-orderings; but we shall not make use of that fact here.

6
Developing mathematics within ZFC

Methods of formalizing parts of mathematics within set theory have developed over a long period, and the methods we present here are not always the first that were given. It is normal nowadays for many of the concepts of mathematics, such as functions, relations, structures, to be given on a set-theoretical basis from the start, and we shall assume this has been done. A systematic collection of definitions for such notions, starting from *ordered pairs*, is given for reference in the appendix chapter 10. Here we give the development of more set-theoretical notions, particularly ordinals and cardinals. We begin with the natural numbers, since that gives a gentler introduction to some of the ideas; and we include a sketch of a possible construction of the real numbers. But our aim is really a development of the ordinal numbers, which are fundamental for the further development of set theory. The cardinal numbers will then be based on the ordinals. Our definitions owe most to von Neumann [vN23].

Note that to be completely formal, we should use only the language introduced in chapter 2 and present all our proofs using the axioms of logic together with the Zermelo–Fraenkel axioms of chapter 3. We shall only approximate this, and add many explanations, so that it may not be trivial to extract from what we give, a completely formal development. It should certainly be possible, but it is not our first aim and we really leave it as an exercise for the interested reader.

6.1 The natural numbers

Of course, we are only giving one possible definition of the natural numbers, and we shall note below what makes this a satisfactory definition. But it is easiest to first give the definitions, and give the comments afterwards.

Definition 6.1.1 $\text{Ind}(X)$ for $\emptyset \in X \land \forall x \in X(x \cup \{x\} \in X)$ (X is an *inductive* set).

This is the fundamental definition, and it introduces the central ideas that we start with $\emptyset$, and that the *successor* of x (i.e. the number one after x, which is usually known as $x+1$) will be taken as $x \cup \{x\}$. So an inductive set is one which has $\emptyset$ as a member, and is closed under successor; i.e. whenever $x \in X$ then also $(x \cup \{x\}) \in X$.

Definition 6.1.2 $\text{Int}(x)$ for $\forall X(\text{Ind}(X) \Rightarrow x \in X)$ (x is a *natural number*, or non-negative *integer*).

$x \dotplus 1$ for $x \cup \{x\}$ (the successor of x; used when x is a natural number).

So a natural number is any set which belongs to every inductive set. Later we shall define general addition of natural numbers, $x+y$, and $x \dotplus 1$ and $x+1$ will be the same thing; but until then we use $\dotplus$ to emphasize that it has a very simple, set-theoretical definition in this context.

As an immediate result of this definition, we can prove the basic property of *mathematical induction*. To make the statement of this more readable, we introduce the restricted variables for natural numbers:

Definition 6.1.3 i, j, k, l, m, n are variables for natural numbers; i.e.

$$\forall i \varphi(i) \quad \text{abbreviates} \quad \forall x(\text{Int}(x) \Rightarrow \varphi(x)), \quad \text{and}$$
$$\exists i \varphi(i) \quad \text{abbreviates} \quad \exists x(\text{Int}(x) \land \varphi(x))$$

(and similarly for the other variables j, k, l, m, n, possibly with suffixes or primes, when we need more).

(Note that there will be occasions in later chapters where we run short of letters for variables, and have to renegue on this definition; it should be clear when this happens.)

Now the theorem which expresses mathematical induction can be given as:

Theorem 6.1.4 $[\varphi(\emptyset) \land \forall n(\varphi(n) \Rightarrow \varphi(n \dotplus 1))] \Rightarrow \forall n \varphi(n)$.

Here φ is any formula of ZF. $\varphi(0)$ is usually called the *basis* of the induction, and $\varphi(n) \Rightarrow \varphi(n \dotplus 1)$ is the *induction step* (which must be proved for all n); within this step, $\varphi(n)$ is the *induction hypothesis*. The conclusion is that $\varphi(n)$ holds for all natural numbers n.

Proof The proof is straightforward from our definitions, but it highlights an important point. We let $X = \{n \mid \varphi(n)\}$ and note from the hypothesis, i.e. from $[\varphi(\emptyset) \land \forall n(\varphi(n) \Rightarrow \varphi(n \dotplus 1))]$, that X is inductive. But every inductive set contains all natural numbers, immediately by 6.1.2. Hence $\forall n \varphi(n)$ as required.

DEVELOPING MATHEMATICS WITHIN ZFC 73

But we must note the important point: we can *only* give the proof as above if we can prove that X is a *set*. In fact we simply cannot apply 6.1.2 unless X is available as a value for our variable X in $\forall X(\mathrm{Ind}(X) \Rightarrow x \in X)$. So before the proof is complete we also need:

Lemma 6.1.5 $\{x \mid \mathrm{Int}(x)\}$ *is a set.*

Proof This is where we make essential use of the axiom of infinity, 3.1.9. (It is in fact the only use we shall make of it directly; every other use will actually be a use of 6.1.5.) In terms of definition 6.1.1, the axiom of infinity says there is a set which is inductive, i.e. $\exists w\,\mathrm{Ind}(w)$. Taking w then as some inductive set, we shall have $\mathrm{Int}(x) \Rightarrow x \in w$ and hence $\{x \mid \mathrm{Int}(x)\} = \{x \mid x \in w \wedge \mathrm{Int}(x)\}$. But $\{x \mid x \in w \wedge \mathrm{Int}(x)\}$ is a set by the subset axiom. □

In view of this, we make the definition:

Definition 6.1.6 ω for $\{x \mid \mathrm{Int}(x)\}$ (the set of natural numbers).

We complete the proof of 6.1.4 by noting that the X required is now $\{n \mid n \in \omega \wedge \varphi(n)\}$ and so is certainly a set. □

In view of 6.1.5 we shall also extend the use of the variables for natural numbers $(i, j, k, l, m, n,$ etc.) to abstraction terms; so we may add to 6.1.3: $\{i \mid \varphi(i)\}$ for $\{x \mid \mathrm{Int}(x) \wedge \varphi(x)\}$. In fact we already did this in the proof of 6.1.4 when we defined X. It is worth noting that 6.1.5 implies that *any* abstraction term of the form $\{i \mid \varphi(i)\}$ (for any formula φ), will be a set (since it is a subset of ω).

Definition 6.1.7 0 for $\emptyset$;
   1 for $\{\emptyset\}$;
   2 for $\{\emptyset, \{\emptyset\}\}$ (i.e. $\{0,1\}$);
   3 for $\{\emptyset, \{\emptyset\}, \{\emptyset, \{\emptyset\}\}\}$ (i.e. $\{0,1,2\}$); etc.
   (Note that $\{\emptyset\}$ is $\emptyset \cup \{\emptyset\}$, i.e. $0 \dotplus 1$; $\{\emptyset, \{\emptyset\}\}$ is $\{\emptyset\} \cup \{\{\emptyset\}\}$, i.e. $1 \dotplus 1$, etc.)

These are the first of the von Neumann natural numbers, and they can be described very simply by the definition: each natural number is the set of all smaller natural numbers. Of course, this is because we know what we mean by smaller; from the set-theoretic side we must *define* what we mean by smaller:

Definition 6.1.8 (i) $x < y$ for $\mathrm{Int}(x) \wedge \mathrm{Int}(y) \wedge x \in y$;
(ii) $x \leq y$ for $x < y \vee x = y$.

So the membership relation is pressed into service as the ordering relation for our natural numbers; and this is implied by our simple description of a natural number as the set of all smaller natural numbers. We still have to prove that it *is* in fact a linear ordering, in other words that it is transitive and antisymmetric, and total on the natural numbers. This last fact is far from obvious from the definition; we need many steps of induction to prove it. First we introduce a useful notion, that of a *transitive set*. This is related to the idea of a transitive *relation*, but is not the same and is easy to confuse if the difference is not pointed out. But the usage is now standard.

Definition 6.1.9 $\mathrm{Trans}(X)$ for $x \in y \wedge y \in X \Rightarrow x \in X$ (X is a *transitive set*). (We have omitted the initial universal quantifiers here, as is common; in full this would be $\forall x \forall y (x \in y \wedge y \in X \Rightarrow x \in X)$. We shall also often abbreviate $x \in y \wedge y \in X$ as $x \in y \in X$.)

Note exercise 6.1.13(1) for many equivalent definitions of $\mathrm{Trans}(X)$; in particular, X is transitive if and only if all members of X are subsets of X (something which is far from true for general sets). The natural numbers we have defined are transitive sets:

Lemma 6.1.10 *(i)* $\mathrm{Int}(i) \Rightarrow \mathrm{Trans}(i)$;
(ii) $\mathrm{Trans}(\omega)$.

Proof (i) By induction on i. We must show $x \in y \in i \Rightarrow x \in i$; for $i = 0$ this is immediate by default since $y \in 0$ is always false. For the induction step, suppose the induction hypothesis (i.e. $x \in y \in i \Rightarrow x \in i$) is true for i, and let $x \in y \in i \dotplus 1$. Then $y \in i \cup \{i\}$, i.e. $y \in i \vee y = i$. In the first case we have $x \in y \in i$ and so $x \in i$ by the induction hypothesis; and in the second case we have $x \in y = i$ which is equivalent to $x \in i$. So in either case we have $x \in i$, and since $i \subseteq i \dotplus 1$ we have proved $x \in y \in i \dotplus 1 \Rightarrow x \in i \dotplus 1$ and completed the induction step. So (i) is proved.

(ii) We must show $x \in y \in \omega \Rightarrow x \in \omega$. We think of this as $x \in i \in \omega \Rightarrow x \in \omega$, and use induction on i. Again for $i = 0$ there is nothing to prove. So we assume the induction hypothesis for i and show $x \in i \dotplus 1 \Rightarrow x \in \omega$. But this is now immediate: if $x \in i \dotplus 1$, then $x \in i \vee x = i$. And if $x \in i$, then $x \in \omega$ by the induction hypothesis; while if $x = i$ then $x \in \omega$ since $i \in \omega$. □

We shall make much use of these properties, eventually without always noting what we are doing. They make the use of the membership relation as the ordering relation on natural numbers much more intuitive; (i) says that the relation is in fact a transitive relation on the natural numbers, and is presumably the source of the slightly confusing terminology. (ii) says that anything counting as less than a natural number must in fact be a natural number.

Before we prove that the ordering relation is total, we show that it is a discrete ordering, with $i \dotplus 1$ as a true successor; in other words nothing can come between i and $i \dotplus 1$:

DEVELOPING MATHEMATICS WITHIN ZFC   75

Lemma 6.1.11 $i < j \Rightarrow i \dotplus 1 \leq j$.

Proof We prove this by induction on j: again for $j = 0$ there is nothing to prove. So assume the statement true for j, and that $i < j \dotplus 1$; we must show $i \dotplus 1 \leq j \dotplus 1$. Since $i \in j \dotplus 1$ we have $i \in j$ or $i = j$; from $i \in j$ by the induction hypothesis we have $i \dotplus 1 \leq j$ (which is $i \dotplus 1 \in j$ or $i \dotplus 1 = j$), and $i = j$ gives $i \dotplus 1 = j \dotplus 1$. Since $j \subseteq j \dotplus 1$, each case gives either $i \dotplus 1 \in j \dotplus 1$ or $i \dotplus 1 = j \dotplus 1$ and hence $i \dotplus 1 \leq j \dotplus 1$ as required. □

Now we are ready to prove that $\in$ is a total ordering of the natural numbers (this is sometimes called the *trichotomy* for natural numbers, since it splits the pairs into three cases):

Theorem 6.1.12 *For any natural numbers i, j, we have $i \in j \vee i = j \vee j \in i$.*

Proof We prove this by induction on j; in other words, we take $\varphi(j)$ to be $\forall i (i \in j \vee i = j \vee j \in i)$, and we prove $\forall j \varphi(j)$ by induction on j. Both the basis and the induction step will require proof by induction on i.

First $\varphi(0)$: for once, the basis of the induction is not quite a triviality. We shall show $\forall i (0 = i \vee 0 \in i)$ (by induction on i). For $i = 0$, $0 = i$ is a triviality. If $0 = i \vee 0 \in i$, then we get $0 \in i \dotplus 1$, as usual immediately from the definition of $i \dotplus 1$.

Finally $\varphi(j \dotplus 1)$, assuming the induction hypothesis $\varphi(j)$. Given i we must show $i \in j \dotplus 1 \vee i = j \dotplus 1 \vee j \dotplus 1 \in i$. But we have assumed $i \in j \vee i = j \vee j \in i$, and $i \in j \vee i = j$ gives $i \in j \dotplus 1$ as we have seen before. And $j \in i$ gives $j \dotplus 1 \leq i$ (by 6.1.11), which is $j \dotplus 1 = i \vee j \dotplus 1 \in i$, and we are through. □

(Note that the final induction on i in the above proof is hidden in the proof of 6.1.11.)

6.1.13 *Exercises*

(1) Show the following are all equivalent to Trans x:

  (i) $\bigcup x \subseteq x$;
  (ii) $y \in x \Rightarrow y \subseteq x$;
  (iii) $x \subset \mathcal{P}(x)$;
  (iv) Trans $\mathcal{P}(x)$;
  (v) $\bigcup (x \cup \{x\}) = x$.

  Show also

  (vi) Trans $x \Rightarrow \emptyset \in x \vee x = \emptyset$ and
  (vii) Trans $x \Rightarrow$ Trans $\bigcup x$; but show the converses fail.

(2) Show that

(i) $x \subseteq i \in \omega \wedge \mathrm{Trans}\, x \Rightarrow x \in \omega$; and hence
(ii) $x \subseteq \omega \wedge \mathrm{Trans}\, x \Rightarrow x \in \omega \vee x = \omega$ and
(iii) $x \subseteq \omega \Rightarrow \bigcup x \in \omega \vee \bigcup x = \omega$.

[Note that these can be proved by induction, or by using the axiom of foundation.]

6.2 The Peano axioms for the natural numbers

Following Dedekind, Peano gave axioms for the natural numbers ([Pea89]). Dedekind had shown that any two models of the natural numbers are isomorphic. For this he used, in effect, the second-order version of the axioms. Here we want to consider the first-order version of these axioms, and we shall call the first-order version PA. Later we shall return to consider the second-order version, and the difference between them, when we present a version of Dedekind's proof in 6.2.13.

All the simple first-order properties of the natural numbers can be derived from PA (though less simple, but still first-order, properties of natural numbers are now known which are provable from ZFC but not from PA, [HP77]). Here we want to show that the natural numbers which we have defined form a model of PA, so that all the simple first-order properties must follow.

First we state the Peano axioms; we give these in the language of arithmetic, which has the operation symbols $+$, $\times$ (binary) and $\dot{+}1$ (unary), and one sort of variable (which we shall write as $i, j, k, \ldots$) to denote natural numbers.

6.2.1 The Peano axioms PA

(i) $0 \neq i \dot{+} 1$
(ii) $i \dot{+} 1 = j \dot{+} 1 \Rightarrow i = j$
(iii) $i + 0 = i$
(iv) $i + (j \dot{+} 1) = (i + j) \dot{+} 1$
(v) $i \times 0 = 0$
(vi) $i \times (j \dot{+} 1) = (i \times j) + i$
(vii) $[\varphi(0) \wedge \forall i(\varphi(i) \Rightarrow \varphi(i \dot{+} 1))] \Rightarrow \forall i \varphi(i)$

In (vii), φ is any formula of the language of arithmetic, and this is the form of the induction axiom for this context. (vii) will follow from 6.1.4 when we give definitions in the language of set theory, for the symbols $+$ and $\times$ of arithmetic, since then we shall be able to translate formulas of arithmetic into formulas of set theory (note that we have already given the definition of $\dot{+}1$ in 6.1.2). Before we do that we check (i) and (ii):

DEVELOPING MATHEMATICS WITHIN ZFC

Lemma 6.2.2 *With the definitions as in 6.1, (i) and (ii) hold.*

Proof (i) $i \dotplus 1 \neq \emptyset$ since $i \in i \dotplus 1$.
(ii) Suppose that $i \dotplus 1 = j \dotplus 1$. Then $i \in i \dotplus 1 = j \dotplus 1$ so $i \in j \cup \{j\}$, i.e. $i \in j$ or $i = j$. Similarly since $j \in j \dotplus 1$ we get $j \in i$ or $j = i$. So if $i \neq j$ we would have both $i \in j$ and $j \in i$, which would contradict the axiom of foundation (see 3.2.3(iii)). □

(An alternative proof of (ii) using induction is in exercise 6.2.15(1). In fact almost all of the development of mathematics we are going to give can be done without using the axiom of foundation, if more sophisticated definitions are used; see 6.5.8(2) and 9.5.8.)

We now give the main result of Dedekind needed to complete the definition of the natural numbers in set theory: we need to show that *recursive definitions* can be carried out. Such definitions occur throughout mathematics, and the simplest form they can have is: first define $f(0)$ outright; then for general i define $f(i+1)$ making use of $f(i)$. Then f is taken to be defined for all natural numbers.

We can regard the Peano axioms (iii) and (iv) as being just such a definition for addition, and then (v) and (vi) as such a definition for multiplication (though these are both functions of two variables, and so not quite the simplest forms of recursive definition; we shall get other forms from the simplest).

The result we prove in ZF is generally known as Dedekind's recursion theorem, although Dedekind gave the result long before ZF was thought of. We give the simplest case, in which the recursion is just *iteration* of the one function f.

Theorem 6.2.3 *(Dedekind's recursion theorem). For any set A, any member $a_0 \in A$, and any function $f : A \to A$, there is a unique function $h : \omega \to A$ satisfying*

(i) $h(0) = a_0$, *and*
(ii) $h(i \dotplus 1) = f(h(i))$ *for all $i \in \omega$.*

Proof We give here a proof which can be described as "building up from below"; another proof is in exercise 6.2.15(2) which can be described as "cutting down from above". The proof "building up from below" has the advantage that it generalizes to the case of *transfinite* recursion, which we introduce in 6.6 below, and that is why we give this proof.

The idea of "building up from below" is very simple: we note that we can start at the bottom (i.e. at 0) and list the values of h as far as we wish. We have immediately $h(0) = a_0$, and then $h(1) = f(a_0)$, then $h(2) = f(f(a_0))$, and so on.

This is clearly why the theorem is accepted in everyday mathematics, often without question or proof; it should be noted that the point of the present proof is not to convince the reader that the theorem is true, but to show that it can be carried out in ZF. As usual, this is a matter of giving a definition in the form of an abstraction term $\{x \mid \varphi(x)\}$ with φ a formula of ZF; and the essential point is to avoid the "and so on" which appears in the description above. So we look for a description of sets which can be gathered together to make up h in one go. That is the point of the next definition: **good** functions are parts of h which start correctly and follow the specification for h (i.e. (i) and (ii)), as far as they go. To be sure that they stay correct (i.e. in agreement with h), we must ensure that we do not allow them to take a value for one number without already having a value for all smaller numbers; this means that the domain of a **good** part must be a transitive set in the sense of 6.1.9 (anything less than a member must already be a member). Natural numbers, i.e. members of ω, are in fact the only proper subsets of ω which are transitive, see exercise 6.1.13(2).

Definition 6.2.4 good g for

$$\mathrm{Func}(g) \wedge \mathrm{dom}(g) \in \omega \wedge \mathrm{range}(g) \subseteq A \wedge$$
$$\wedge [0 \in \mathrm{dom}(g) \Rightarrow g(0) = a_0] \wedge$$
$$\wedge \forall i[i \dotplus 1 \in \mathrm{dom}(g) \Rightarrow g(i \dotplus 1) = f(g(i))].$$

Some examples: $\{\langle 0, a_0\rangle\}$ is **good** (with domain 1);
$\{\langle 0, a_0\rangle, \langle 1, f(a_0)\rangle\}$ is **good** (with domain 2);
$\{\langle 0, a_0\rangle, \langle 1, f(a_0)\rangle, \langle 2, f(f(a_0))\rangle\}$ is **good** (with domain 3);
and we could clearly continue ad inf. Note that $\emptyset$ will also count as **good** (with domain 0).

The idea now is to set $\Gamma = \{g \mid \mathbf{good}\, g\}$ and $h = \bigcup \Gamma$, so that h has as members all those ordered pairs that occur as members of some **good** g.

Let us first check that these will be *sets*. We have specified that **good** $g \Rightarrow \mathrm{dom}(g) \subseteq \omega \wedge \mathrm{range}(g) \subseteq A$, so that we have **good** $g \Rightarrow g \subseteq \omega \times A$. (Here $\times$ is the cartesian product, see 10.2.1.) So members of Γ are subsets of $\omega \times A$, i.e. $\Gamma \subseteq \mathcal{P}(\omega \times A)$ and so Γ is a set by the subset and power-set axioms, which can also be used to prove that the cartesian product of sets is a set, and by the axiom of infinity, which was used to show that ω is a set. (Note that we have automatically assumed A is a set from the start, since it is denoted by a variable.)

Now since h is given as $\bigcup \Gamma$, h will also be a set by the union axiom. (In fact we could avoid introducing Γ by defining h directly as $\{\langle i, a\rangle \mid \exists g[\mathbf{good}\, g \wedge i \in \mathrm{dom}(g) \wedge g(i) = a]\}$, then $h \subseteq \omega \times A$ and h is shown to be a set without using the power-set axiom, as in exercise 6.2.15(2).)

Now to complete the proof of 6.2.3, we need two lemmas:

DEVELOPING MATHEMATICS WITHIN ZFC

Lemma 6.2.5 **good** *functions cannot disagree; that is,*

$$[\mathbf{good}\, g \wedge \mathbf{good}\, g' \wedge i \in \mathrm{dom}(g) \cap \mathrm{dom}(g')] \Rightarrow g(i) = g'(i).$$

Proof By induction on i. It is clear for $i = 0$ since $g(0) = g'(0) = a_0$ (provided $0 \in \mathrm{dom}(g) \cap \mathrm{dom}(g')$). And if $i \dotplus 1 \in \mathrm{dom}(g) \cap \mathrm{dom}(g')$, then also $i \in \mathrm{dom}(g) \cap \mathrm{dom}(g')$ (since both are transitive; this is where we use that fact). So $g(i) = g'(i)$ by the induction hypothesis, and hence

$$g(i \dotplus 1) = f(g(i)) = f(g'(i)) = g'(i \dotplus 1)$$

as required. □

Lemma 6.2.6 *There are enough* **good** *functions; that is, for all natural numbers i there is a* **good** *function g with $i \in \mathrm{dom}(g)$.*

Proof Again by induction on i (how else?). The example $\{\langle 0, a_0 \rangle\}$ was given above, and has 0 in its domain. And suppose that g is **good** with $i \in \mathrm{dom}(g)$ but $i \dotplus 1 \notin \mathrm{dom}(g)$. Then we can define $g' = g \cup \{\langle i \dotplus 1, f(g(i)) \rangle\}$. It is straightforward to check that g' is **good** (using $g'(i) = g(i)$), and $i \dotplus 1 \in \mathrm{dom}(g')$, as required. □

From these two lemmas and the definition of h given above, we have immediately that h is a function (by 6.2.5), and that $\mathrm{dom}(h) = \omega$ (by 6.2.6). h will satisfy (i) and (ii) by the definition of **good**, and the uniqueness of h is by another application of 6.2.5. So theorem 6.2.3 is proved. □

For future use, we note that we can extract from the above proof a formula $H(A, a_0, f, i, y)$ with the property that whenever A, a_0, and f satisfy the hypotheses of 6.2.3 then $H(A, a_0, f, i, y)$ will hold if and only if $h(i) = y$, see exercise 6.2.15(3). And in fact we can omit mention of the variable A, since $A = \mathrm{dom}(f)$ and so can be recovered from f.

6.2.7 Applications of the recursion theorem

We made informal uses of the recursion theorem in 4.2.6, but we shall leave the formalization of that to the exercise 6.2.15(4). The first applications we give are to the definitions of addition and multiplication for natural numbers. As noted above, these are functions of two variables, but we can vary the starting value a_0 as the second variable here. This is equivalent to defining many functions of one variable, and then we put them together again:

Definition 6.2.8 For each $m \in \omega$, set $p_m : \omega \to \omega$ to be the function given by

$$p_m(0) = m,$$
$$p_m(i \dotplus 1) = p_m(i) \dotplus 1$$

and set $q_m : \omega \to \omega$ to be the function given by

$$q_m(0) = 0,$$
$$q_m(i \dotplus 1) = p_m(q_m(i)).$$

Then let $i + j = p_i(j)$ and $i \times j = q_i(j)$.

Lemma 6.2.9 *The functions $+$ and $\times$ satisfy the Peano axioms 6.2.1(iii)–(vi).*

Proof This is because the functions p_m and q_m will satisfy their defining equations, by theorem 6.2.3, and these were chosen to be the Peano axioms as required. □

In fact the alert reader should notice that we have cheated a little here: the defining equations for multiplication in 6.2.8 are not *exactly* the Peano axioms for multiplication as given in 6.2.3(v), (vi) since (merely for convenience of printing and reading) in 6.2.8 the definition of q_m gives $m \times (i \dotplus 1) = m + (m \times i)$, while the Peano axiom (vi) was $m \times (i \dotplus 1) = (m \times i) + m$. So we have to prove the commutativity of addition before 6.2.9 is fully proved. We shall, of course, leave that as an exercise, see 6.2.15(5).

6.2.10   Remark

The formal proof that addition and multiplication can be defined in ZF needs to use the remark at the end of 6.2.6 to show that we can give formulas of ZF, $P(i, j, k)$ and $Q(i, j, k)$, for which we can prove that for any $i, j, k \in \omega$, $P(i, j, k)$ holds if and only if $p_i(j) = k$, and $Q(i, j, k)$ holds if and only if $q_i(j) = k$. Only when this is done can we regard the proof that the axiom of induction holds for our model, as complete. Details are left to the reader, see exercise 6.2.15(3). Then we can, if we wish, regard $+$ and $\times$ as themselves being sets, by defining $+$ as $\{\langle\langle i, j\rangle, k\rangle \mid P(i, j, k)\}$ and $\times$ as $\{\langle\langle i, j\rangle, k\rangle \mid Q(i, j, k)\}$.

Note that this now gives us three different ways to think of $+$ and $\times$: first, as formal symbols of the language of arithmetic; second, as operations on numbers; third, as certain sets of triples of members of ω as we have just suggested. The first way of thinking gives no meaning to the symbols unless we add the Peano axioms (or some other equivalent axioms) and regard the meaning of the symbols to be given in some sense by the theorems which can then be derived. The second way is the naive way that we all learned, probably first in primary school; and the claim that mathematics is included in set theory could be regarded as the identification of the second with the third way above. A less extreme way to regard the matter would be to note that the third way can be taken as a model, or explanation, of the second.

DEVELOPING MATHEMATICS WITHIN ZFC   81

However there is a further point to note when we realize that we have made set theory itself into a formal system, ZF, and the definition we have given above makes + and × into defined terms of ZF. Then these again can only be regarded as having the meaning given to them by virtue of the theorems which can be derived in ZF or ZFC. In fact we know we can prove *more* theorems about them in ZF than in PA (see [HP77]), and this illustrates yet another sense in which mathematics can be done in set theory.

6.2.11   The uniqueness of the natural numbers

Dedekind used his recursion theorem to prove that the natural numbers are unique, and we shall give a version of this proof, as a further application. It also serves as a very clear example of the difference between first-order and second-order logic, as introduced in 2.4. We give the proof first in a naive way, and then look at how it can be interpreted. We need the general notion of a *Peano system*:

Definition 6.2.12 A *Peano system* is a structure $\langle N, \mathbb{O}, S \rangle$ where $\mathbb{O} \in N$ and $S : N \to N$, and

(i) $\mathbb{O} \neq S(n)$ for any $n \in N$;
(ii) $S(n) = S(n') \Rightarrow n = n'$ for any $n, n' \in N$; and
(iii) for any set $Z \subseteq N$, if $\mathbb{O} \in Z$ and for all $n \in N$, $n \in Z \Rightarrow S(n) \in Z$, then $Z = N$.

These are just three of the Peano axioms, with a set form for the induction axiom; and Dedekind's intention for this was clearly that it should be taken in the second-order sense: (iii) should hold for *every* subset Z of N. But we can only write a first-order statement in LST, and theorem 6.1.4 shows that the structure $\langle \omega, 0, \dot{+}1 \rangle$ is a Peano system in this first-order sense. Dedekind showed that Peano systems are unique up to isomorphism; we shall show:

Theorem 6.2.13 *For any Peano system $\langle N, \mathbb{O}, S \rangle$ there is a unique isomorphism h from $\langle \omega, 0, \dot{+}1 \rangle$ onto $\langle N, \mathbb{O}, S \rangle$.*

Proof   We use the recursion theorem to define $h : \omega \to N$ by:

$$h(0) = \mathbb{O},$$
$$h(i \dot{+} 1) = S(h(i)).$$

This definition immediately makes h preserve the structure. Then the proof that h is one-to-one is by induction and is straightforward, using properties (i) and (ii) of the fact that $\langle N, \mathbb{O}, S \rangle$ is a Dedekind system. We prove that h is onto N:

Let $Z = \mathrm{range}(h)$. Then $\mathbb{O} \in Z$ since $\mathbb{O} = h(0)$. And if $n \in Z$, let $n = h(i)$. Then $h(i \dotplus 1) = S(h(i)) = S(n)$ and so $S(n) \in Z$. Hence by the induction axiom (iii) for the Dedekind system $\langle N, \mathbb{O}, S \rangle$, $Z = N$ and h is onto. □

6.2.14 Remark

As noted, this result needs to be interpreted with care in view of the limitations of first-order logic (of 2.4.6). We know that any formal system for arithmetic must have models which are not isomorphic; how does that fit with the above proof? Peano's axioms, if taken with the definitions we have given in ZFC, give a formal, first-order system of arithmetic, and so will have non-isomorphic models.

But Dedekind's axiom 6.2.11(iii) is clearly intended to be taken in the second-order sense; *every* subset of N should satisfy it. And in this sense, although we can't set out a complete proof system for second-order logic, it seems reasonable (and agreed by most mathematicians) that the steps required to prove Dedekind's result (just as in the proof of 6.2.13), are correct and hence any two structures which are second-order Peano structures (i.e. which satisfy the second-order sense of 6.2.11(iii)), will be isomorphic. This is certainly the sense in which the result was understood, when first proved. This second-order notion of a Peano structure is what is usually referred to as the *standard* integers. Within the cumulative type structure, our definition of ω will give a Peano structure in this second-order sense (since within the cumulative type structure all subsets are present), and so the cumulative type structure contains a model of the standard integers.

However, second-order logic is not formal, and it is worth noting what can be said about this result of Dedekind, when we restrict ourselves to formal methods, i.e. to first-order logic. PA as given in 6.2.1 is a set of first-order axioms, and if 6.2.11 is considered formally, as stated in LST, then we have a first-order notion of a Peano system. (Within a model of ZF, the LST version of 6.2.11(iii) will hold for a structure if it holds for the subsets of N present in the model.)

In this sense, what we proved above can be thought of as proving that, within any one model of ZF set theory, all Peano systems, and hence all models of arithmetic which, within that model, satisfy the set form of induction, i.e. 6.2.11(iii) rather than just 6.2.1(vii), will be isomorphic by isomorphisms which are within that model. But the limitations of first-order logic tell us that there must be more than one model of ZF set theory, and the models of arithmetic within these different models of ZF can also be different (i.e. non-isomorphic).

But we can say more, since Gödel's incompleteness theorem can in fact be stated and proved within ZF itself. So we can use that to show that, within any one model of ZF, there will be models of PA which are not in fact Peano systems (so that although they satisfy all instances of the induction axiom which can be given using formulas of PA, i.e. 6.2.1(vii), there will be some sets, not given by formulas of PA, for which induction would fail; they would not satisfy 6.2.11(iii)). These latter are called non-standard models of PA, and they will exist within any model of ZF. The uniqueness we have proved shows that each model of ZF will have its own version of a standard model of PA, (that one isomorphic to ω in that model of ZF), and would justify calling it *the standard* model of PA (within that model of ZF). But another model of ZF may have a different standard model of PA.

Many mathematicians will not want to restrict themselves to these formal considerations, and will want to say that the cumulative type structure is *the standard model* for ZF, and that the standard model for PA within *that* is the *genuine* standard model of arithmetic (and also the genuine ω). This is sometimes called the *platonist* or *realist* position on the foundations of mathematics, and seems to be the most natural way to think, if we are to have a single, unique standard model of arithmetic. See 9.6 for further remarks about these matters.

Some further properties of the natural numbers ω are in the exercises, but we may in fact assume from now on any standard property we need as a theorem of ZF (since its proof from the Peano axioms will now be available as a proof in ZF). We shall need several in the constructions of the rational and real numbers, which we sketch in the next sections.

6.2.15   Exercises

(1) Prove Peano axiom (ii), i.e. $i \dot{+} 1 = j \dot{+} 1 \Rightarrow i = j$, without using the axiom of foundation.
   [Use induction instead. Prove $i < j \Rightarrow j \not\leq i$ using 6.1.11.]

(2) Prove 6.2.3 (Dedekind's recursion theorem) by "cutting down from above" as follows:
   For $X \subseteq \omega \times A$, say X is **full** if $\langle 0, a_0 \rangle \in X$ and $\langle j, c \rangle \in X \Rightarrow \langle j+1, f(c) \rangle \in X$. Define h by
   $$\langle j, c \rangle \in h \Leftrightarrow \forall X (X \text{ \textbf{full}} \Rightarrow \langle j, c \rangle \in X).$$
   Show that h must be a function with domain ω and that h will satisfy the recursion equations.
   [Use induction on j. Note the parallel here with the definition of inductive sets in 6.1.1; indeed a more general form for **full** could be used, not restricting to subsets of $\omega \times A$, but allowing arbitrary inductive sets as the domains for **full** sets.]

(3) Write the formula $H(A, a_0, f, i, y)$ as in the note after the proof of the recursion theorem, so that whenever A, a_0 and f satisfy the appropriate properties, $H(A, a_0, f, i, y)$ will hold if and only if $h(i) = y$.

[Two possible forms for H are:

$$H(A, a_0, f, i, y) \Leftrightarrow \exists g(g \text{ \bf good} \wedge i \in \text{dom } g \wedge g(i) = y); \text{ and}$$
$$H(A, a_0, f, i, y) \Leftrightarrow \forall g(g \text{ \bf good} \wedge i \in \text{dom } g \Rightarrow g(i) = y).$$

Prove that these are equivalent. Other forms can be given from the proof in exercise (2) above.]

Now write formulas $P(i, j, k)$ and $Q(i, j, k)$ as required in 6.2.10 to complete the proof that addition and multiplication are available in ZF.

(4) Show that the proof of the Schröder–Bernstein theorem 4.2.5 can be presented in ZF.

[Use the recursion theorem to give the required definitions, which were given naively in chapter 4.]

(5) Prove that addition and multiplication on ω are associative and commutative, and also distributive, and that the cancellation laws hold. Also show that the monotonicity laws hold for $+$ and $\times$ with respect to $<$, and that subtraction holds in the form:

$$i \leq j \Leftrightarrow \exists k(i + k = j).$$

[These are all proved by induction; it seems easiest first to prove associativity for addition, then commutativity, beginning with $0 + n = n + 0$ and $1 + n = n + 1$, then distributivity, $n(m + p) = nm + np$, then associativity of multiplication, with commutativity of multiplication last. In most cases, induction on the rightmost variable will be needed. After some steps, such exercises become exercises in arithmetic, and there is no obvious end to them; elementary number theory is the name usually given to what can be deduced from the Peano axioms by first-order logic, and of course it is not a decidable collection.]

(6) Show that the natural numbers can serve as finite cardinals: show that $m \sim n \Leftrightarrow m = n$ for members of ω.

[Show by induction that $m \sim n \Rightarrow m = n$; at the induction step, use $m \neq 0 \Rightarrow \exists k(m = k + 1)$ and construct a mapping $k \sim n$ from a mapping $m \sim n + 1$.]

(7) The *transitive closure* of a set. We give this construction here since it is very close to an application of the recursion theorem; but since it makes essential use of the replacement axiom, it has to involve more than just the simple version of the recursion theorem we have given.

The problem is simple: show that for every set x, there is a transitive set to which x belongs.

[Informally, show that $\text{TC}(x) = \{x\} \cup x \cup \bigcup x \cup \bigcup \bigcup x \cup \ldots$ is a transitive set to which x belongs, and is the smallest such set.

So we need to show that this is a set, and the recursion involved is simple enough: we want to define

$$h(0) = \{x\}; \quad h(n+1) = \bigcup h(n)$$

and then $\mathrm{TC}(x)$ will be $\bigcup\{h(n) \mid n \in \omega\}$.

Since we are not given, from the start, a set A which contains all the values of h as members, it is easiest to remake the proof of the recursion theorem, building in the required proof that the resulting collection is a set using replacement on ω. Replace the definition of **good** by the following: let $F(n, x, y)$ be

$$\exists f(\mathrm{dom}(f) = n + 1 \wedge f(0) = x \wedge f(n) = y \wedge \forall i < n(f(i+1) \in f(i))).$$

Show by induction on n that $\{y \mid F(n, x, y)\}$ is a set, for each n. Let $G(n) = \{y \mid F(n, x, y)\}$ and show that $G(n+1) = \bigcup G(n)$ (noting that the pair-set axiom is used here to prove the existence of the needed functions f). So $G(n)$ is the required $h(n)$.

Next, let $X = \{G(n) \mid n \in \omega\}$ and use the replacement axiom to show that X is a set; then show that $\mathrm{TC}(x) = \bigcup X$.]

6.3 The rational numbers

First we give the construction of the integers, that is the positive and negative integers $\mathbb{Z}$. These will be defined using pairs of natural numbers with the intention that the pair $\langle m, n \rangle$ should represent the integer $m - n$. Each integer will have many representations, and we will therefore have an equivalence relation which will hold between pairs which represent the same integer. So we immediately know the equivalence relation required:

$\langle m, n \rangle$ is equivalent to $\langle m', n' \rangle$ if and only if $m - n = m' - n'$.

But this is circular, and no good as a definition yet. If $m < n$ then $m - n$ is one of the negative integers we have yet to introduce! But elementary arithmetic gives us a simple transformation: since

$$m - n = m' - n' \quad \text{if and only if} \quad m + n' = m' + n$$

we can give the definition in ZF as:

Definition 6.3.1 $\sim_{\mathbb{Z}}$ is the equivalence relation on $\omega \times \omega$ given by:

$$\langle m, n \rangle \sim_{\mathbb{Z}} \langle m', n' \rangle \quad \text{if and only if} \quad m + n' = m' + n$$

and $\mathbb{Z}$ (the set of all positive and negative integers) is the set of equivalence classes of $\omega \times \omega$ under $\sim_{\mathbb{Z}}$.

So an integer will be a set such as $[\langle m,n\rangle]_{\sim_\mathbb{Z}}$, which is the equivalence class of $\langle m,n\rangle$ under $\sim_\mathbb{Z}$, i.e. the set $\{\langle m',n'\rangle \in \omega \times \omega \mid \langle m',n'\rangle \sim_\mathbb{Z} \langle m,n\rangle\}$. (We write $[x]_\sim$ for the equivalence class of x under equivalence relation $\sim$, and then we shall omit the suffix $\sim$ whenever possible.) Clearly all of this can be carried out using the definitions which we have already given for ω, and the operation $+$ on ω, in ZF. We need:

Proposition 6.3.2 $\sim_\mathbb{Z}$ *is an equivalence relation.*

We leave this as an exercise, 6.3.9(1); it is a matter of proving properties of $+$ on ω, which follow from the Peano axioms, rather than a matter of set theory.

Of course, $\mathbb{Z}$ should have operations and an ordering relation:

Definition 6.3.3 (i) $[\langle m,n\rangle] +_\mathbb{Z} [\langle m',n'\rangle]$ is the equivalence class

$$[\langle m+m', n+n'\rangle]$$

(where $+$ is the operation of addition on ω already defined).
(ii) $[\langle m,n\rangle] \times_\mathbb{Z} [\langle m',n'\rangle]$ is the equivalence class

$$[\langle mm'+nn', mn'+m'n\rangle]$$

(where we have reverted to the common notation of juxtaposition rather than $\times$, so that mm' stands for $(m \times m')$, and this operation is multiplication on ω, already defined).
(iii) $-_\mathbb{Z}[\langle m,n\rangle]$ is the equivalence class $[\langle n,m\rangle]$ (and note that no operation on ω is involved here).
(iv) The relation $<_\mathbb{Z}$ is given by: $[\langle m,n\rangle] <_\mathbb{Z} [\langle m',n'\rangle]$ if and only if $m+n' < m'+n$ (where $<$ is the relation on ω already defined).

Of course, there should be no mystery about these definitions: all can be obtained very simply by taking the known, intended properties of $\mathbb{Z}$ such as: $(m-n)+(m'-n')=(m+m')-(n+n')$, and rewriting terms such as $m-n$ as $[\langle m,n\rangle]$. But there is an important point about all such definitions for equivalence classes. Since they are given in terms of *representatives* of the equivalence classes, and each class has many (infinitely many in this case) possible choices of representatives, we must confirm that we get the same answer from our definition, whichever representatives are chosen.

This is clearer if we give an example. -2 is, in our model of $\mathbb{Z}$, the equivalence class $[\langle 0,2\rangle]$; it is also $[\langle 1,3\rangle]$ and $[\langle 2,4\rangle]$ and $[\langle 13,15\rangle]$ etc., since these are all exactly the same equivalence class. So our definition of $-2 +_\mathbb{Z} -2$ will give $[\langle 0,4\rangle]$ and $[\langle 2,6\rangle]$ and $[\langle 4,8\rangle]$ and $[\langle 26,30\rangle]$ and many other possibilities. We note that these are, in fact, all the *same* equivalence class, and this is essential if $+_\mathbb{Z}$ is to be well-defined, with *one* value and not many, for each pair of arguments. So we must prove the following:

Proposition 6.3.4 *The definitions in 6.3.3 are all well-defined; that is, if $\langle m, n \rangle \sim_{\mathbb{Z}} \langle i, j \rangle$ and $\langle m', n' \rangle \sim_{\mathbb{Z}} \langle i', j' \rangle$, then also*
in the case of (i): $\langle m + m', n + n' \rangle \sim_{\mathbb{Z}} \langle i + i', j + j' \rangle$;
and for (ii): $\langle mm' + nn', mn' + m'n \rangle \sim_{\mathbb{Z}} \langle ii' + jj', ij' + i'j \rangle$;
and for (iii): $\langle n, m \rangle \sim_{\mathbb{Z}} \langle j, i \rangle$;
and for (iv): $m + n' < m' + n$ *if and only if* $i + j' < i' + j$.

Proof See exercise 6.3.9(1).

6.3.5 Remark

We can now go on and establish further properties of our model of the integers, see exercise 6.3.9(2). But one property is clearly missing: we do *not* get the natural numbers as a subset of the integers. The best we can do is to *embed* ω into $\mathbb{Z}$ by the mapping $n \mapsto n_{\mathbb{Z}} = [\langle n, 0 \rangle]$. This will clearly have all the right properties, in the sense that it will carry the operations and the ordering from ω to $\mathbb{Z}$, and its range, the non-negative integers $\mathbb{Z}^{\geq 0}$, is isomorphic to ω. But it is clearly not *natural* to regard the number 1 as the set $\{\emptyset\}$ when it is thought of as a natural number, and as the set $[\langle 1, 0 \rangle]$, i.e. $\{\langle 1, 0 \rangle, \langle 2, 1 \rangle, \langle 3, 2 \rangle, \ldots\}$, when it is thought of as an integer.

There are various ways of mitigating this *unnaturalness*, and we give two of these below. But none are convincing, in the sense that we would not expect mathematicians in general to accept them as in any sense *true*. This shows that what we are doing in this chapter is best regarded as showing ways in which mathematics can be modelled within set theory, and so providing a possible foundation, rather than as showing what numbers etc. really *are*. This would accord with the view that mathematics is about structures only up to isomorphism, and the point of our current work is to show that the basic structures (integers, real numbers), have models within ZF.

6.3.6 Less unnatural models

The simplest way to produce a model of the integers which contains the natural numbers is to replace the natural numbers; in effect we change our mind about the natural numbers and no longer take ω as our model of the natural numbers, but the image of ω under the embedding into $\mathbb{Z}$. We might call this image $\mathbb{N}$, and we certainly have $\mathbb{N} \subset \mathbb{Z}$. A disadvantage now is that we shall be constructing the rational numbers $\mathbb{Q}$ and then the real numbers $\mathbb{R}$, and at each step just the same problem occurs; if we change our mind each time about what the previous numbers are, we end with four versions (and a fifth if we go on to construct the complex numbers $\mathbb{C}$).

An alternative way is to change the model $\mathbb{Z}$ that we have produced by removing the image of ω and sticking in its place the original ω. If we set

$\mathbb{Z}^- = \{[\langle m,n \rangle] \mid m < n\}$ then the new model is $\mathbb{Z}^* = \omega \cup \mathbb{Z}^-$, and new operations $+$, $\times$, etc. must be defined on $\mathbb{Z}^*$. These will now extend the operations defined originally on ω, but the detail of the definitions will be quite clumsy. One way to define $+_{\mathbb{Z}^*}$, for example, would be:

Set $h : \omega \to \mathbb{Z}$ to be the embedding $h(n) = [\langle n, 0 \rangle]$ and then extend h to $h^* : \mathbb{Z}^* \to \mathbb{Z}$ by taking $h^* = h$ on ω and the identity on $\mathbb{Z}^-$. Then for any $p, q \in \mathbb{Z}^*$, define $p +_{\mathbb{Z}^*} q = h^{*-1}(h^*(p) +_{\mathbb{Z}} h^*(q))$. Similar definitions can be given for the other operations and relations.

The obvious advantage of this procedure is that if we repeat it at each successive stage, as we construct $\mathbb{Q}$, $\mathbb{R}$ and $\mathbb{C}$, then we can end up with definitions of $\mathbb{Q}^*$, $\mathbb{R}^*$ and $\mathbb{C}^*$ which satisfy $\omega \subset \mathbb{Z}^* \subset \mathbb{Q}^* \subset \mathbb{R}^* \subset \mathbb{C}^*$, and the number 1 will mean the same whether considered as a natural number or as a complex number. But the unnaturalness stands out when we consider how different will be the number 1 and the number 1.41 and the number $\sqrt{2}$ when considered as sets.

For the rest of our presentation in this section, we shall ignore this problem; we shall give constructions of the rationals $\mathbb{Q}$ and the reals $\mathbb{R}$ for which we do not have either $\mathbb{Z} \subset \mathbb{Q}$ or $\mathbb{Q} \subset \mathbb{R}$. We shall only have embeddings, and we shall continue to use $\mathbb{Z}$ as constructed in 6.3.1.

The construction of the rationals $\mathbb{Q}$ from the integers $\mathbb{Z}$ is perhaps the best known of these steps, using "vulgar fractions". We simply give the definitions, leaving the details as exercises:

Definition 6.3.7 $\sim_\mathbb{Q}$ is the equivalence relation on $\mathbb{Z} \times \mathbb{Z}'$ (where $\mathbb{Z}'$ is the non-zero integers $\mathbb{Z} - \{0_\mathbb{Z}\}$), given by

$$\langle p, q \rangle \sim_\mathbb{Q} \langle p', q' \rangle \text{ if and only if } pq' = p'q.$$

Here the operation $\times_\mathbb{Z}$ on the right is being written as juxtaposition, as is common.

$\mathbb{Q}$ is the set of equivalence classes of $\mathbb{Z} \times \mathbb{Z}'$ under $\sim_\mathbb{Q}$.

The operations on $\mathbb{Q}$ and the ordering relation are given by:

(i) $[\langle p, q \rangle] +_\mathbb{Q} [\langle p', q' \rangle]$ is the equivalence class $[\langle pq' + p'q, qq' \rangle]$ (where now all equivalence classes are taken with respect to $\sim_\mathbb{Q}$, and the operations which are used on the right will always be those of $\mathbb{Z}$).
(ii) $[\langle p, q \rangle] \times_\mathbb{Q} [\langle p', q' \rangle]$ is the equivalence class $[\langle pp', qq' \rangle]$.
(iii) $-_\mathbb{Q}[\langle p, q \rangle]$ is the equivalence class $[\langle -p, q \rangle]$.
(iv) Provided $p \neq 0_\mathbb{Z}$, $[\langle p, q \rangle]^{-1}$ is the equivalence class $[\langle q, p \rangle]$ (and this is not defined if $p = 0_\mathbb{Z}$).
(v) $[\langle p, q \rangle] <_\mathbb{Q} [\langle p', q' \rangle]$ if and only if $pq' <_\mathbb{Z} p'q$.

All of this is simply the usual rules for fractions, with $[\langle p, q \rangle]$ replacing the notation $\frac{p}{q}$. It even accords with common practice in that equivalent fractions

DEVELOPING MATHEMATICS WITHIN ZFC

are regarded as equal. Historically this must have been the first example of an equivalence relation, and of taking equivalence classes; all others must have followed this example. For us, however, there is further work to be done, to prove that all the appropriate properties can be proved in ZF.

Proposition 6.3.8 *The relation $\sim_{\mathbb{Q}}$ is an equivalence relation, and the definitions given in 6.3.7 for the operations and relations of $\mathbb{Q}$ are independent of the choice of representatives from the equivalence classes involved.*

Proof   See exercise 6.3.9(1) □

The natural embedding of $\mathbb{Z}$ into $\mathbb{Q}$ is now given by $p \mapsto [\langle p, 1\rangle]$ for $p \in \mathbb{Z}$. This will carry $\mathbb{Z}$ to an isomorphic copy in $\mathbb{Q}$, and as noted above it is the only natural thing we can do; there are only unnatural ways to have a model with $\mathbb{Z} \subset \mathbb{Q}$.

6.3.9   Exercises

(1) Show that $\sim_{\mathbb{Z}}$ is an equivalence relation, and that the operations on $\mathbb{Z}$ are well-defined for this equivalence relation (i.e. prove the results set out in 6.3.4).

[These constitute further properties of addition and multiplication of natural numbers, and all should follow from the properties in 6.2.15(5).]

Also prove that $\sim_{\mathbb{Q}}$ is an equivalence relation, and that the operations on $\mathbb{Q}$ are well-defined for this equivalence relation (i.e. prove the results in 6.3.8).

(2) Prove that addition and multiplication in $\mathbb{Z}$ are associative and commutative and distributive, and further that $\mathbb{Z}$ has no zero divisors. Then prove that $\mathbb{Q}$ is a field, and that it has the Archimedean property: for any $r \in \mathbb{Q}$ there is a finite number k such that $r <_{\mathbb{Q}} 1_{\mathbb{Q}} + 1_{\mathbb{Q}} + \ldots + 1_{\mathbb{Q}}$, the sum taken k times.

[Of course for this last property we take finite numbers as being members of ω, which makes it a triviality.]

6.4   The real numbers

We shall give Dedekind's construction here. Cantor's construction used Cauchy sequences and is indicated in exercise 6.4.7(2), as is the use of decimal expansions. Dedekind based his construction on the notion of a *cut* or *section* in the rationals: a cut being a split of the rationals into a lower and an upper part, with the intention that it should represent the unique real number which lies between the two parts (in the intuitive picture of the real numbers as the *number line*). For simplicity we take just the lower, or left, half of the split as being sufficient, and we define:

Definition 6.4.1 A *left Dedekind section* is a subset X of the rationals $\mathbb{Q}$ satisfying:

(i) $\emptyset \neq X \neq \mathbb{Q}$ (so X is a non-empty, proper subset);
(ii) if r, s are rationals with $r <_\mathbb{Q} s$ and $s \in X$ then $r \in X$ (this says that X is a *left-section* of $\mathbb{Q}$);
(iii) X has no greatest member, i.e. if $r \in X$ then there is some $s \in X$ with $r <_\mathbb{Q} s$.

A *real number* is a left Dedekind section (of the rationals, as above).
$\mathbb{R}$ is the set of all left Dedekind sections.

Note that $\mathbb{R} \subset \mathcal{P}(\mathbb{Q})$ and so is a set, by the power-set axiom. This is one use of the power-set axiom that cannot be circumvented, as we shall note in 6.7.8(3).

We shall show the main properties of $\mathbb{R}$; first we give the embedding of $\mathbb{Q}$ into $\mathbb{R}$:

Definition 6.4.2 The natural embedding $h : \mathbb{Q} \to \mathbb{R}$ is given by
$$h(r) = r^* = \{s \in \mathbb{Q} \mid s <_\mathbb{Q} r\}.$$

Here we can see the point of (iii) in definition 6.4.1. There are two natural candidates for left parts of the cut in the rationals which should correspond to a rational number r, according to whether we allow r itself to belong to the left side of the cut, or the right. We have chosen to leave it on the right, so that no left Dedekind section will have a greatest member.

This particular technicality is the only messy point about this definition of the real numbers; it has to be considered in the definition of negatives and inverses. Multiplication has also to be given by cases for positives and negatives. But addition, and the ordering relation, are very natural:

Definition 6.4.3 For $X, Y \in \mathbb{R}$ we define:

(i) $X +_\mathbb{R} Y = \{r +_\mathbb{Q} s \mid r \in X \land s \in Y\}$.
(ii) $X <_\mathbb{R} Y$ if and only if $X \subset Y$.
(iii) $-_\mathbb{R} X = \{s \in \mathbb{Q} \mid s <_\mathbb{Q} -_\mathbb{Q} r \text{ for some } r \notin X\}$.
(iv) $\mathbb{R}^+$ for $\{X \in \mathbb{R} \mid 0_\mathbb{Q} \in X\}$ (the positive reals).
(v) If $X, Y \in \mathbb{R}^+$ then $X \times_\mathbb{R} Y = \{r \times_\mathbb{Q} s \mid 0 \leq_\mathbb{Q} r \in X \land 0 \leq_\mathbb{Q} s \in Y\} \cup 0^*$ (here 0^* is the set of all negative rationals, which is also the zero real $0_\mathbb{R}$, and must be a subset of each positive real).
(vi) $0_\mathbb{R} \times_\mathbb{R} X = X \times_\mathbb{R} 0_\mathbb{R} = 0_\mathbb{R}$ for all reals X;
   if $X \in \mathbb{R}^+$ and $Y \notin \mathbb{R}^+$ then $X \times_\mathbb{R} Y = Y \times_\mathbb{R} X = -_\mathbb{R}(X \times_\mathbb{R} (-_\mathbb{R} Y))$;
   if $X, Y \notin \mathbb{R}^+$ then $X \times_\mathbb{R} Y = (-_\mathbb{R} X) \times_\mathbb{R} (-_\mathbb{R} Y)$.

DEVELOPING MATHEMATICS WITHIN ZFC   91

(vii) If $X \in \mathbb{R}^+$ then $X^{-1} = \{s \in \mathbb{Q} \mid \exists r \notin X (s <_{\mathbb{Q}} r^{-1})\}$;
if $-_{\mathbb{R}} X \in \mathbb{R}^+$ then $X^{-1} = -_{\mathbb{R}} (-_{\mathbb{R}} X)^{-1}$; and $0_{\mathbb{R}}^{-1}$ is not defined.

Note that we have not tried to label the inverse operator $^{-1}$ with a suffix $\mathbb{R}$ or $\mathbb{Q}$ as appropriate, but the reader should by now be able to tell which is being defined and which is the operator for rationals as given in 6.3.7.

Proposition 6.4.4 *With these operations and ordering relation, $\mathbb{R}$ is a real-closed ordered field, with the Archimedean property; further, $\mathbb{R}$ is order-complete.*

Proof We leave most of this to the exercise 6.4.7(1), and comment only on the closure and the Archimedean property. Here the *Archimedean property* is essentially the property that every positive real is below some natural number, though this is usually expressed by saying it is below a sum of the form $1_{\mathbb{R}} +_{\mathbb{R}} 1_{\mathbb{R}} +_{\mathbb{R}} \ldots +_{\mathbb{R}} 1_{\mathbb{R}}$ for a finite number of $1_{\mathbb{R}}$'s. Since we have defined *finite* in terms of our natural numbers, we need only show that the embedding of the natural numbers into $\mathbb{R}$ (via $\mathbb{Z}$ and $\mathbb{Q}$) is cofinal in $\mathbb{R}$.

It should be clear that the embedding of ω into $\mathbb{Z}$ was cofinal, and similarly for the embedding of $\mathbb{Z}$ into $\mathbb{Q}$. Now if $X \in \mathbb{R}$ then X is a proper subset of $\mathbb{Q}$, and if $r \in \mathbb{Q} - X$ then $X \leq_{\mathbb{R}} r^*$. If $r = [\langle p, q \rangle]$ with p positive, then p is (the image of) a natural number, which will be (after embedding into $\mathbb{Q}$) $\geq r$, and hence (after embedding into $\mathbb{R}$) $\geq X$. If p is negative then we can use $-_{\mathbb{Z}} p$.

To say that $\mathbb{R}$ is *real-closed* means that every polynomial of odd degree has a root in $\mathbb{R}$. Historically this was the motivation for extending the rationals to the reals, so that irrationals such as $\sqrt{2}$ were included. Since it follows from the order-completeness, we show that. In view of the very natural definition of the ordering relation in this model, it is not surprising that the *supremum* is also very simple:

Lemma 6.4.5 *If $X \subset \mathbb{R}$ is a non-empty bounded set of reals, then $\sup X = \bigcup X$.*

Proof This is a standard property of any set ordered by $\subseteq$, provided $\bigcup X$ is in the set; see example 5.2.6(v). So we just have to check that $\bigcup X$ is a real number, i.e. is a left Dedekind section, when X is non-empty and bounded above. That is left as an exercise, 6.4.7(1). □

6.4.6   Uniqueness of $\mathbb{R}$

Just as we can show that the natural numbers are unique up to isomorphism, we can now show that the real numbers are unique up to isomorphism. Many descriptions of $\mathbb{R}$ can be given. One is that $\mathbb{R}$ is an ordered field which is

Archimedean and order-complete; another is that it is the order-completion of the rationals; there are others. For each of these descriptions there is a proof that any structure fitting that description is isomorphic to the set $\mathbb{R}$ defined in 6.4.1. Some of these are indicated in 6.4.7(4).

As in 6.2.14 in connection with the natural numbers, this fact of uniqueness must be regarded with care. It will have different meaning according to whether we work in second-order logic (where the quantifiers involved in "Archimedean" and "order-complete" really mean "for all subsets of the reals"), or in first-order logic (where they can only refer to subsets which occur in a model). It is only in the second-order sense that we really have a unique model of the reals (up to isomorphism). This notion, *the standard real numbers,* can be justified by working within the cumulative type structure, the genuine or intended model of ZF.

But, in the first-order sense, it can only mean that within each model of ZF there is (up to isomorphism) just one model of the real numbers, which will be the standard reals for that model. Non-standard models can arise when axioms such as the Archimedean property and the order-completeness axiom are restricted to a first-order sense. (See 9.6 for further remarks on these questions.)

6.4.7 *Exercises*

(1) Show that if X is a non-empty bounded set of reals, then $\bigcup X$ is a real and is $\sup X$. Hence complete the proof that $\mathbb{R}$ is a real-closed Archimedean ordered field.

(2) Cantor's construction of $\mathbb{R}$. Cantor gave the following construction of the real numbers from the rationals:

Say that a sequence of rationals (a_n) (i.e. a function from ω into $\mathbb{Q}$) is a *Cauchy* sequence if it satisfies the general principle of convergence as given by Cauchy, i.e.

$$\forall \varepsilon > 0 \exists N \forall m, n > N (|a_m - a_n| < \varepsilon$$

(this all interpreted as in $\mathbb{Q}$, i.e. $\varepsilon \in \mathbb{Q}$).

For any two such Cauchy sequences of rationals (a_n), (b_n), say $(a_n) \sim (b_n)$ if and only if $(a_n - b_n) \to 0$, i.e. is a null sequence. Show that this is an equivalence relation on Cauchy sequences of rationals, and take $\mathbb{R}$ as the set of equivalence classes under this relation. $\mathbb{Q}$ will be embedded in $\mathbb{R}$ via the constant sequences, i.e. $r \mapsto [(a_n)]$ where $a_n = r$ for all n.

If $[(a_n)]$ and $[(b_n)]$ are two such equivalence classes of Cauchy sequences of rationals, define $[(a_n)] +_\mathbb{R} [(b_n)]$ and $[(a_n)] \times_\mathbb{R} [(b_n)]$ as $[(a_n +_\mathbb{Q} b_n)]$ and $[(a_n \times_\mathbb{Q} b_n)]$ respectively. Show that these are well-defined, and that they make $\mathbb{R}$ into a real-closed ordered field which is order-complete.

[In this context it is most natural to prove completeness in the form:

every Cauchy sequence of reals converges to a real; or equivalently, every monotone bounded sequence of reals converges to a real.]
(3) Decimal representations of reals. Show that each real defined as above, i.e. each equivalence class of Cauchy sequences of rationals, will contain at least one representative which is a decimal expansion, i.e. a sequence (a_n) of rationals such that each a_n is of the form $a_n = a_0 + b_n \cdot 10^{-n}$ with $a_0 \in \mathbb{Z}$, $b_0 = 0$, $b_n \in \omega$, and satisfying $10 b_n \le b_{n+1} \le 10 b_n + 9$ for each n. Hence show that the usual decimal representation of reals gives a structure isomorphic to that given by the Cauchy sequences of reals.
(4) Uniqueness of the reals. Show that any dense linear ordering without endpoints, which is order-complete and has a countable dense subset, is isomorphic to the order-type of the reals.

[The first step is to show that any two countable dense linear orderings without endpoints are isomorphic. Let $\langle A, <_1 \rangle$ and $\langle B, <_2 \rangle$ be the given countable orderings; the isomorphism can be constructed step by step from given enumerations (a_n), (b_n) of A and B, working alternately mapping the first element a_n not yet used in A to the first element of B which stands in the same relation under $<_2$ to each previously used element of B, as a_n stands under $<_1$ to all the previously used elements of A. Then switch and do the same for the first element b_m not yet used in B. In this way a mapping is built which will be a partial isomorphism at each stage, and the final result will be an isomorphism between the whole of A and the whole of B. (This form of construction was given by Cantor and is sometimes described as Cantor's "back and forth argument".)

To complete the proof, show that each cut in the countable dense subset must correspond to exactly one element of the given order-complete ordering, so that this must be isomorphic to the structure which would result from the construction by Dedekind sections.]

Hence show that the constructions of Cantor (in (2) above, via Cauchy sequences) and the construction using decimal expansions (in (3) above) are isomorphic to the construction using Dedekind sections in 6.4.1.

6.5 Ordinals in ZF: basic properties

We did not give any development of the informal notion of *ordinal* in chapter 5, but historically it was developed informally. Modern development usually follows the definitions given by von Neumann (also used earlier by Mirimanoff, though not developed to the same extent; see [Hal84]). We have already used this for the natural numbers in 6.1. Although ordinals are a direct extension of the natural numbers, it is simplest to use a completely new definition, and only later will we see that it is in fact an extension of the definitions in 6.1. We repeat the definition of transitive sets from 6.1.9:

Definition 6.5.1 (i) Trans x for $y \in z \wedge z \in x \Rightarrow y \in x$ (x is *transitive*);
(ii) Connex x for $y \in x \wedge z \in x \Rightarrow y \in z \vee y = z \vee z \in y$ (x is *connected under* $\in$);
(iii) Ord x for Trans $x \wedge$ Connex x (x is an *ordinal*);
(iv) $x < y$ for Ord $x \wedge$ Ord $y \wedge x \in y$ (x is a smaller ordinal than y);
(v) $x \leq y$ for Ord $x \wedge$ Ord $y \wedge (x < y \vee x = y)$.
(von Neumann's definition of ordinals).

So an ordinal is a transitive set which is linearly ordered by the membership relation. The definition of $x < y$ immediately identifies *smaller ordinals* with *members* (of an ordinal), and repeats the slogan from 6.1.7 usually used to describe the von Neumann ordinals: Each ordinal is the set of all smaller ordinals. But we must of course prove that the members of an ordinal are in fact ordinals. This relies heavily on the axiom of foundation:

Proposition 6.5.2 *(i)* Ord $x \wedge y \in x \Rightarrow$ Ord y
(ii) Ord $\emptyset$
(iii) Ord $x \Rightarrow$ Ord$(x \cup \{x\})$
(iv) Int $x \Rightarrow$ Ord x
(v) Ord ω.

Proof For (i) we are given Trans x and Connex x and $y \in x$ and we must show Trans y and Connex y. It is immediate to see Connex y follows from these. But for Trans y we need the axiom of foundation: given $v \in y$ and $u \in v$, we must show $u \in y$. By Trans x and Connex x we have $u \in y \vee u = y \vee y \in u$, and the axiom of foundation rules out both $u = y$ and $y \in u$. (Compare exercise 3.2.3(3).) (ii) and (iii) are straightforward. For (iv) and (v), we have Trans by 6.1.10, and Connex by 6.1.12. □

This makes our ordinals, extensions of the natural numbers. As in 6.1 the most important property remaining to prove is the *trichotomy* for ordinals:

Theorem 6.5.3 Ord $x \wedge$ Ord $y \Rightarrow x \in y \vee x = y \vee y \in x$.

Proof Rather than using induction, as in the proof for the natural numbers, we make use of the axiom of foundation, using the lemma:

Lemma 6.5.4 *If x and y are ordinals and $x \subset y$ with $x \neq y$, then $x \in y$.*

Proof Take $Y = y - x$ and apply the axiom of foundation. $Y \neq \emptyset$ by the assumptions, so we get $v \in Y$ with $v \cap Y = \emptyset$. We show that $v = x$.
So suppose $u \in v$; then $u \in y$ (by Trans y) but $u \notin Y$ (since $v \cap Y = \emptyset$), hence $u \in x$, i.e. $v \subseteq x$.

Next suppose $u \in x$; then $u \in y$ and $v \in y$, so by Connex y, $u \in v \vee u = v \vee v \in u$. But $u = v \vee v \in u$ implies $v \in x$ (since Trans x), contradicting $v \in Y$. So we must have $u \in v$, i.e. we have $x \subseteq v$.
Hence $v = x$ as claimed. □

Now to complete the proof of 6.5.3, we consider $x \cap y$. It is immediate from the definitions of Trans and Connex that $\text{Ord}\, x \wedge \text{Ord}\, y \Rightarrow \text{Ord}(x \cap y)$, so we apply 6.5.4 to $x \cap y$. If we assume $x \cap y \neq x$ and $x \cap y \neq y$ then we get both $x \cap y \in x$ and $x \cap y \in y$, i.e. $x \cap y \in x \cap y$, which again contradicts the axiom of foundation. So we must have either $x \cap y = x$ or $x \cap y = y$; that is, either $x \subseteq y$ or $y \subseteq x$. Applying 6.5.4 again gives $x \in y \vee x = y \vee y \in x$ as required. □

This result really completes the definition of ordinals, in the sense that we now have all the basic properties available. We continue with some more standard ideas; we introduce the usual relativized variables:

Definition 6.5.5 We use lower-case greek letters α, β, γ, ξ, η, ζ, etc. (but not φ, ψ, χ) as variables for ordinals, so that:

$$\forall \alpha \varphi(\alpha) \text{ abbreviates } \forall \alpha(\text{Ord}(\alpha) \Rightarrow \varphi(\alpha))$$
$$\exists \alpha \varphi(\alpha) \text{ abbreviates } \exists \alpha(\text{Ord}(\alpha) \wedge \varphi(\alpha))$$
$$\{\alpha \mid \varphi(\alpha)\} \text{ abbreviates } \{\alpha \mid \text{Ord}(\alpha) \wedge \varphi(\alpha)\}.$$

And we note the main distinction between the two types of ordinals:

Definition 6.5.6 $\text{Suc}\, x$ for $\text{Ord}\, x \wedge \exists y(x = y + 1)$ (x is a *successor* ordinal);
$\text{Lim}\, x$ for $\text{Ord}\, x \wedge \neg \text{Suc}\, x \wedge x \neq 0$ (x is a *limit* ordinal).

We show in exercise 6.5.8(1) some of the reasons which make the word *limit* appropriate here. Some writers include 0 as a limit ordinal.

6.5.7 The Burali-Forti paradox

As with all the set-theoretic paradoxes, this is for us simply the observation that a certain collection (in this case the collection of all ordinals) is a proper class and not a set. Let $\text{On} = \{x \mid \text{Ord}\, x\}$, and assume this is a set. Then $\text{Trans}(\text{On})$ and $\text{Connex}(\text{On})$ by 6.5.2(i) and 6.5.3; i.e. $\text{Ord}(\text{On})$. But then $\text{On} \in \text{On}$, contradicting the axiom of foundation. Hence On must be a proper class.

(The paradox still occurs if we give definitions which do not make use of the axiom of foundation, as in exercise (2) below.)

This completes the presentation of the early paradoxes, which was begun in 1.3.

6.5.8 Exercises

(1) Show:

   (i) $\operatorname{Int} x \Leftrightarrow (x = 0 \vee \operatorname{Suc} x) \wedge \forall y < x(y = 0 \vee \operatorname{Suc} y)$
   (ii) $\operatorname{Lim} x \Leftrightarrow \operatorname{Ord} x \wedge x \neq 0 \wedge \forall y \in x(y+1 \in x)$
   (iii) $\operatorname{Lim} \omega$
   (iv) $\forall x \in X(\operatorname{Ord} x) \Rightarrow \operatorname{Ord} \bigcup X$
   (v) $\forall x \in X(\operatorname{Ord} x) \wedge X \neq \emptyset \Rightarrow \operatorname{Ord} \bigcap X \wedge \bigcap X \in X$.

   Show further that if X is a set of ordinals, then $\bigcup X$ is the supremum of X in the linear ordering of the ordinals, and $\bigcap X$ is the infimum, and hence any set of ordinals is well-ordered by $\in$.

   Show also that if $f : \omega \to \operatorname{On}$ and is increasing, then $\bigcup_{n \in \omega} f(n)$ is a limit ordinal; similarly for any limit ordinal λ if $g : \lambda \to \operatorname{On}$ is increasing, then $\operatorname{Lim}(\bigcup_{\alpha < \lambda} g(\alpha))$.

(2) Alternative definitions of ordinals. Show that the following is equivalent to the definition in 6.5.1:

   $\operatorname{Ord}' x$ for $\operatorname{Trans} x \wedge \forall y \in x(\operatorname{Trans} y)$.

   [$\operatorname{Ord} x \Rightarrow \operatorname{Ord}' x$ follows from 6.5.2(i). $\operatorname{Ord}' x \Rightarrow \operatorname{Ord} x$ also requires the axiom of foundation: suppose $\operatorname{Ord}' x \wedge \neg \operatorname{Connex} x$, and let $y_0 \in Y = \{y \in x \mid \exists z \in x \neg(y \in z \vee y = z \vee z \in y)\}$ be minimal, i.e. so that $y_0 \cap Y = \emptyset$. Then let $y_1 \in Y_1 = \{z \in x \mid \neg(y_0 \in z \vee y_0 = z \vee z \in y_0)\}$ be minimal, so that $y_1 \cap Y_1 = \emptyset$. Then show that $y_1 \subseteq y_0$ and $y_0 \subseteq y_1$, giving a contradiction.]

   Show that a further definition equivalent to 6.5.1 is:

   $\operatorname{Ord}'' x$ for $\operatorname{Trans} x \wedge (\in \operatorname{wo} x)$, where $(\in \operatorname{wo} x)$ is

   $$\forall u \subseteq x(u \neq \emptyset \Rightarrow \exists v \in u \forall w \in u(v \in w \vee v = w)),$$

   i.e. x is well-ordered by the membership relation.

   Show that $\operatorname{Ord}'' x \Rightarrow \operatorname{Ord} x$ without any use of the axiom of foundation, and further that all the properties of 6.5.2 to 6.5.4, and exercise (1) above, and the Burali-Forti paradox, hold for Ord'' (again without any use of the axiom of foundation).

   [Now the Burali-Forti paradox is simplest in the form of noting that if On is a set, then not only $\operatorname{On} \in \operatorname{On}$ but also $\operatorname{On} \dotplus 1 \in \operatorname{On}$ and the pair $\{\operatorname{On}, \operatorname{On} \dotplus 1\}$ contradicts the well-ordering of On by $\in$, since it has no first member.

   This definition is probably closest to von Neumann's original definition, and has been used by authors who wished to avoid appeal to the axiom of foundation. It is straightforward to check that, using it, mathematics can be developed within set theory without any use of the axiom of foundation. Within set theory we shall give uses of the axiom of foundation which are essential; and we shall also show in 9.5.8 that we can use this sort of definition to give a model of set theory including the axiom of foundation, within the theory without that axiom, and hence show the *relative consistency* of the axiom of foundation.]

6.6 Transfinite induction

The presentation of ordinals we have given, relying on the axiom of foundation, does not stress what was historically the central idea of the ordinals: that they are *well-ordered* by their ordering relation. If we wish to develop the theory without relying on the axiom of foundation, we have to recast the definition, so that besides transitivity and connectedness, we also say that the membership relation is well-founded on an ordinal (as in exercise 6.5.8(2) above). The axiom of foundation says this is so for any set, and makes our definitions that much simpler.

As already noted in 5.4, the main property of well-orderings is that it allows us to use induction. When we pass beyond the finite, we call this *transfinite induction*, following Cantor.

Transfinite induction works over any well-founded relation (see 6.6.13(5)), but we shall begin with transfinite induction over the ordinals. First we give the theorem which extends mathematical induction:

Theorem 6.6.1 $\forall \alpha [(\forall \beta < \alpha \varphi(\beta)) \Rightarrow \varphi(\alpha)] \Rightarrow \forall \alpha \varphi(\alpha)$.

Proof Assume $\neg \forall \alpha \varphi(\alpha)$, say $\neg \varphi(\xi)$ for an ordinal ξ. Then consider the set $X = \{\gamma \mid \gamma \leq \xi \wedge \neg \varphi(\gamma)\}$. $X \subseteq \xi + 1$ and so is a set. And $\xi \in X$ so $X \neq \emptyset$; so by the axiom of foundation there is $\gamma \in X$ with $\gamma \cap X = \emptyset$. But then γ contradicts the hypothesis; $\forall \beta < \gamma \varphi(\beta)$ will hold, but $\varphi(\gamma)$ fails. □

Note that this is more like "course of values" recursion (or induction), than the simplest form; we are assuming as induction hypothesis $\forall \beta < \alpha \varphi(\beta)$, rather than just $\varphi(\beta)$ for the previous value β. This is forced upon us, since for limit ordinals there is no unique "previous" value; see 6.5.8(1).

Now we must consider the theorem which extends the recursion theorem; this is sometimes called "definition by transfinite induction". There is a complication caused by the fact that On is a proper class and not a set. We have to introduce some variant notations if we want the theorem and its proof to look anything like the simpler recursion theorem (in its "course of values" form, just as above). (Another way of handling this was given by von Neumann; see 9.2.)

First the intention: we want to say that if F is any function at all (which is going to tell us how to get from previous values to the next value, of the function H which is to be defined by the recursion), then there will be a (unique) function H, which will satisfy the recursion equations. What should these equations look like? Instead of $h(n+1) = f(h(n))$, as in the simple recursion theorem, we must have $H(\alpha) = F(H \restriction \alpha)$, where $H \restriction \alpha$ must code up the previous values of H. We do this by:

Definition 6.6.2 $H \restriction \alpha$ for $\{\langle \beta, H(\beta) \rangle \mid \beta < \alpha\}$ (so that this is the function H restricted to the set α, which is what we want, since members of α are the ordinals less than α).

Now if F is suitably clever, it should be able to use any information it wants about any of the previous values of H. But we still haven't looked at the problem of dealing with proper classes. We really want to say that F can be any function at all, from V to V, where V is the universe; and we shall want the resulting H to be a function from On to V. So F and H can't be sets either, and we shall proceed by giving suitable meanings to formulas such as $H(\alpha) = F(H \restriction \alpha)$, when F and H are not variables but themselves *formulas* (in fact any formulas).

Definition 6.6.3 If $F(x, x_1, \ldots, x_n, y)$ is any formula with at least x and y free, then $F'x$ is given by

$$F'x = \{z \mid \exists y [\forall w (F(x, x_1, \ldots, x_n, w) \Leftrightarrow w = y) \wedge z \in y]\}.$$

Here $x_1, \ldots, x_n$ are just parameters, but x is being treated as an argument and y as a value; and if, for any x, there is a unique y satisfying $F(x, x_1, \ldots, x_n, y)$, then $F'x$ will be that y; if not, it will just be the empty set. (We are borrowing the notation F' which Russell used for function values, see 10.2.3(iii).)

Now, if we cheat and write $F(x)$ instead of $F'x$ (which we have already done with $H(\beta)$ in definition 6.6.2), we can give the theorem we want as:

Theorem 6.6.4 *For any formula* $F(x, x_1, \ldots, x_n, y)$, *there is a formula* $H(x, x_1, \ldots, x_n, y)$ *such that for all* α, $H(\alpha) = F(H \restriction \alpha)$.

The proof of this is parallel to the proof of the recursion theorem in 6.2.3. First we note one fact: the axiom of replacement is essential in proving this. In fact we can see one place where this is needed, as soon as we look at the justification of $H \restriction \alpha$. This will in general have one member for each member of α, but these will not come from any given set, and the only way to justify this as a set is to use replacement (taking α as the starting set and replacing each β from α by $\langle \beta, H(\beta) \rangle$).

Proof We deliberately use the same word in our temporary definition as we did in the proof of the recursion theorem, and define **good** by:

DEVELOPING MATHEMATICS WITHIN ZFC

Definition 6.6.5 g **good** for

$$\operatorname{Func} g \wedge \operatorname{Ord}(\operatorname{dom} g) \wedge \forall \alpha \in \operatorname{dom} g[g(\alpha) = F(g \upharpoonright \alpha)].$$

This is parallel to the definition in 6.2.3. Here we want the domain to be an ordinal, so that we can go further than the integers (which is all we were interested in in 6.2.3), but the reason is the same: to be a good part of the intended result, we must ensure that previous values are right, before we can expect the next value to be right. And ordinals are transitive, so that the previous values will be there, if the domain is an ordinal. We only need one clause to say how the values are given by previous values, since we are taking the "course of values" form of recursion.

We can now prove two lemmas, parallel to 6.2.5 and 6.2.6:

Lemma 6.6.6 good *functions cannot disagree; i.e. if g* **good** *and g'* **good** *and $\alpha \in \operatorname{dom} g \cap \operatorname{dom} g'$ then $g(\alpha) = g'(\alpha)$.*

Proof By induction on α (now transfinite induction): the induction hypothesis will be

$$[\beta < \alpha \wedge \beta \in \operatorname{dom} g \cap \operatorname{dom} g'] \Rightarrow g(\beta) = g'(\beta).$$

But since the domains are ordinals, and so transitive,

$$\beta < \alpha \Rightarrow [\beta \in \operatorname{dom} g \cap \operatorname{dom} g'],$$

and hence $g \upharpoonright \alpha = g' \upharpoonright \alpha$. Hence

$$g(\alpha) = F(g \upharpoonright \alpha) = F(g' \upharpoonright \alpha) = g'(\alpha),$$

as required. □

Lemma 6.6.7 *For all ordinals α, there is a* **good** *function g with $\alpha \in \operatorname{dom} g$.*

Proof By induction on α; now the induction hypothesis is

$$\forall \beta < \alpha \exists y[y \text{ good} \wedge \beta \in \operatorname{dom} g],$$

Here we must do more work than in 6.2.6. We cannot just take a previous value; we must construct the function which takes all previous values. Let

$$G = \{\langle \beta, g(\beta)\rangle \mid \beta < \alpha \wedge g \text{ good} \wedge \beta \in \operatorname{dom} g\}.$$

We show G is a set by replacement on α: for each $\beta \in \alpha$ we get at most one pair $\langle \beta, g(\beta) \rangle$ by 6.6.6. Also we get at least one such pair by the induction hypothesis. Hence also G is a function with domain α, and for each $\beta < \alpha$, $G \restriction \beta = g \restriction \beta$ for some (indeed, any) **good** function g with $\beta \in \operatorname{dom} g$. So $G(\beta) = F(G \restriction \beta)$, and G is **good**, with $\operatorname{dom} G = \alpha$. Now set

$$G' = G \cup \{\langle \alpha, F(G) \rangle\},$$

and we have the required result, since it is now immediate that G' is **good** and $\alpha \in \operatorname{dom} G'$. □

Now to complete the proof of 6.6.4, we actually have less to do than we had in the proof of the recursion theorem: we simply define the formula $H(x, x_1, \ldots, x_n, y)$ to be

$$\exists g [g \textbf{ good} \wedge x \in \operatorname{dom} g \wedge g(x) = y].$$

Note that the parameters $x_1, \ldots, x_n$, if any, are present, though hidden, in the definition of **good**. The two lemmas give all we need to complete the proof. Indeed, it is worth noting that we can prove more: we have also

$$\forall x [\operatorname{Ord} x \Rightarrow \exists^1 y H(x, x_1, \ldots, x_n, y)]$$

and that the definition of H is equivalent to

$$\forall g [g \textbf{ good} \wedge x \in \operatorname{dom} g \Rightarrow g(x) = y].$$

□

6.6.8 Ordinal arithmetic

Just as the first use of the recursion theorem was to give definitions in set theory for the arithmetic functions (addition, multiplication and exponentiation), we can now use transfinite recursion to extend these operations to all ordinals. We use the definitions:

$$\alpha + 0 = \alpha, \quad \alpha + (\beta + 1) = (\alpha + \beta) + 1, \quad \text{and}$$
$$\alpha + \lambda = \bigcup_{\beta < \lambda} (\alpha + \beta) \quad \text{if } \operatorname{Lim} \lambda;$$

$$\alpha \times 0 = 0, \quad \alpha \times (\beta + 1) = (\alpha \times \beta) + \alpha, \quad \text{and}$$
$$\alpha \times \lambda = \bigcup_{\beta < \lambda} (\alpha \times \beta) \quad \text{if } \operatorname{Lim} \lambda;$$

$$\alpha^0 = 1, \quad \alpha^{(\beta+1)} = \alpha^\beta \times \alpha, \quad \text{and}$$
$$\alpha^\lambda = \bigcup_{\beta < \lambda} \alpha^\beta \quad \text{if } \operatorname{Lim} \lambda.$$

DEVELOPING MATHEMATICS WITHIN ZFC

The clauses for limit ordinals are the only changes from the definitions for the operations on the natural numbers in 6.2.8. So these are clearly extensions of those operations into the transfinite. But the properties do not all extend: e.g.

6.6.9 Examples

(i) Since $\operatorname{Lim}\omega$, we get $1 + \omega = \bigcup_{n \in \omega} 1 + n = \omega \neq \omega + 1$ and so ordinal addition is not commutative.

(ii) Similarly $2 \times \omega = \bigcup_{n \in \omega} 2 \times n = \omega \neq \omega + \omega = \omega \times 2$ and ordinal multiplication is not commutative.

(iii) $2^\omega = \bigcup_{n \in \omega} 2^n = \omega$, but $\omega^2 = \omega \times \omega = \bigcup_{n \in \omega} \omega \times n$ and this is much greater that ω; indeed, $\omega \times 2 = \omega + \omega = \bigcup_{n \in \omega} \omega + n$ is the next limit ordinal after ω, and $\omega \times 3 = (\omega + \omega) + \omega = \bigcup_{n \in \omega} (\omega + \omega) + n$ is the next after that, and so on. So $\omega \times \omega$ is the first ordinal which is a limit of limit ordinals.

(Note that this form of exponentiation is *ordinal exponentiation*, and is a very different operation to the *cardinal exponentiation* of 4.3. It corresponds to the restricted product in 5.4.6(3) and the form of exponentiation in 5.4.6(5). This correspondence is in exercise 6.6.13(2).)

In contrast with commutativity, we can show that ordinal addition and multiplication are still *associative*, and also distributive on one side (but not the other). These, and properties of exponentiation, are in the exercise 6.6.13(1).

6.6.10 Order-types of well-orderings

We noted in 5.4 that the original intention of ordinals was to provide suitable representatives of isomorphism classes of well-ordered sets. Transfinite recursion is needed to show that we do get this property:

Theorem 6.6.11 *For every well-ordered poset $\langle X, < \rangle$ there is a unique ordinal α and unique isomorphism $h : \alpha \to X$.*

Proof The definition of h by transfinite recursion would be straightforward: it has to be

$$h(\beta) = x, \quad \text{where } x \text{ is the } < \text{-least in } X - \{h(\gamma) \mid \gamma < \beta\},$$

except that this does not work when $X - \{h(\gamma) \mid \gamma < \beta\} = \emptyset$. To make the definition fit the theorem we have proved, we use a stratagem (which will be used often): let z_0 be some fixed element with $z_0 \notin X$. Now define

$$H(\beta) = x, \quad \text{where } x \text{ is the } \prec\text{-least in } X - \{H(\gamma) \mid \gamma < \beta\}$$
$$\text{unless } X - \{H(\gamma) \mid \gamma < \beta\} = \emptyset, \text{ when } H(\beta) = z_0.$$

Then let α be the first ordinal for which $H(\alpha) = z_0$, and define $h = H \restriction \alpha$. We need to show that there will be such a first ordinal α, then the definition of α will ensure that $h : \alpha \to X$ and that h is onto X. Writing $h''\beta$ as usual for $\{h(\gamma) \mid \gamma < \beta\}$, we will have for $\gamma < \beta < \alpha$ that $h(\gamma)$ is $\prec$-least in $X - h''\gamma \supset X - h''\beta$, and $h(\beta) \in X - h''\beta$, $h(\gamma) \notin X - h''\beta$. So $h(\gamma) \prec h(\beta)$ and $h(\gamma) \neq h(\beta)$; that is, h is one–one and order-preserving, and hence an isomorphism as required.

So we must show there is such an ordinal; we need the axiom of replacement. Suppose no such α; then for every ordinal α, $H(\alpha)$ will be defined and will be a member of X, and the argument above shows H will be a one–one function from all ordinals into X. But now that implies that the collection of all ordinals is a set, since X is (by assumption); the axiom of replacement is applied, with formula $\psi(x, y)$ as

$$\text{Ord } y \wedge H(y) = x$$

and we get that On $= \{y \mid \exists x \in X \psi(x, y)\}$, where On is the class of all ordinals, and was shown to be a proper class in 6.5.8.

Hence the required α must exist. $\square$

Corollary 6.6.12 *Any two well-orderings are comparable; that is, if $\langle X, \mathbf{r} \rangle$ and $\langle Y, \mathbf{s} \rangle$ are both well-ordered posets, then there is an order-preserving embedding from $\langle X, \mathbf{r} \rangle$ into (or onto) $\langle Y, \mathbf{s} \rangle$, or vice versa.*

Proof Using 6.5.3, this is true for the ordinals isomorphic to $\langle X, \mathbf{r} \rangle$ and $\langle Y, \mathbf{s} \rangle$. $\square$

6.6.13   Exercises

(1) Prove that addition and multiplication of ordinals are associative, and distributive in that $\alpha(\beta + \gamma) = \alpha\beta + \alpha\gamma$.

Show also that $\alpha^\beta \alpha^\gamma = \alpha^{\beta+\gamma}$ and $(\alpha^\beta)^\gamma = \alpha^{\beta\gamma}$ for ordinal exponentiation; but show that $(\alpha\beta)^\gamma = \alpha^\gamma \beta^\gamma$ can fail.

Show that one-sided cancellation holds for addition, multiplication and ordinal exponentiation: if $\alpha + \beta = \alpha + \gamma$ then $\beta = \gamma$; if $\alpha \geq 1$ and $\alpha\beta = \alpha\gamma$ then $\beta = \gamma$; and if $\alpha \geq 2$ and $\alpha^\beta = \alpha^\gamma$ then $\beta = \gamma$.

(2) Show that the definitions of addition, multiplication and exponentiation for well-orderings given in 5.4.6 coincide with the definitions in 6.6.8.

[It is straightforward to see that the inductive clauses used in 6.6.8 will hold for the constructions given in 5.4.6, so an inductive proof follows.]

(3) Other uses of transfinite induction. Definition of the cumulative hierarchy and *ranks*.

Define $V_0 = \emptyset$, $V_{\alpha+1} = \mathcal{P}(V_\alpha)$, and $V_\lambda = \bigcup_{\alpha<\lambda} V_\alpha$ for λ a limit ordinal. Show that every set is a member of some V_α.

[First, using the replacement axiom, show: if each member of x is in some V_α, then there is some β such that x is a subset of V_β. Hence $x \in V_{\beta+1}$. Then use the axiom of foundation.]

Define rank $x = \alpha$ if $x \subseteq V_\alpha$ but $\neg x \subseteq V_\beta$ for all $\beta < \alpha$. Show that $x \in V_\alpha \Leftrightarrow \operatorname{rank} x < \alpha$, and that rank x satisfies the recursion on $\in$ given by

$$\operatorname{rank} x = \bigcup \{\operatorname{rank} y + 1 \mid y \in x\}.$$

Show further that V_ω is the set of all hereditarily finite sets (introduced in 3.2.3(4)); and that the same hierarchy is obtained if we start with $V_0 = \{\emptyset\}$ instead of $V_0 = \emptyset$ (with finite ranks reduced by one).

[This shows that the cumulative hierarchy is not changed if we consider $\emptyset$ to be the only urelement, as mentioned in 1.4.2.]

(4) Recursion over general well-founded relations. The definition of rank in (3) above is a typical definition by recursion over the membership relation. Such definitions can be given over any *well-founded* relation (such as the membership relation); two properties are needed for such definitions to work. One is the usual idea of the relation being well-founded, namely that every non-empty subset of the domain should have *minimal* members (under the relation involved; in contrast with well-ordered relations, these will not usually be unique). Under the axiom of choice, this property is equivalent to there being no infinite descending sequences under the relation.

The other property needed is a closure property: the collection of "smaller elements" (i.e. preceding elements under the relation), should always be a *set*. For the membership relation this is immediate; but note that the use made of it will be to show that any set can be extended to a larger set which is *closed downwards* under the relation. (Such closed downwards sets will be the appropriate domains for the **good** functions, in this context.) The construction of such closed downwards sets is done exactly parallel to the construction of the transitive closure of a set 6.2.15(7) (which gives a set closed downwards under the membership relation, i.e. transitive in the sense of definition 6.1.9).

Let $\varphi(x,y)$ be any formula with x and y free, and define Wf $\varphi(x,y)$ for

$$\forall X [X \neq \emptyset \Rightarrow \exists u \in X \forall v \in X \neg \varphi(v,u)] \wedge \forall u \exists Y \forall v [\varphi(v,u) \Rightarrow v \in Y].$$

Show that if Wf $\varphi(x,y)$ then every set X can be extended to a set Y which is φ-closed, i.e. show that

$$\forall X \exists Y [X \subseteq Y \wedge \forall u, v (u \in Y \wedge \varphi(v,u) \Rightarrow v \in Y)].$$

[This can be done by a construction parallel to the construction of the transitive closure of a set.]

Then show that recursion can be carried out over the relation $\varphi(x, y)$, if Wf $\varphi(x, y)$; i.e. show that given any formula $F(x, x_1, \ldots, x_n, y)$ there is a formula $H(x, x_1, \ldots, x_n, y)$ such that if Wf $\varphi(x, y)$, then for all x, $H(x) = F(H \restriction \varphi^{-1}(x))$, where $\varphi^{-1}(x)$ is $\{u \mid \varphi(u, x)\}$ and $H \restriction \varphi^{-1}(x)$ is $\{\langle u, H(u) \rangle \mid \varphi(u, x)\}$.

One major use of such recursion is to define a notion of rank corresponding to any well-founded relation (just as rank was defined for membership in (3) above); show that if Wf $\varphi(x, y)$ then we can define a notion $\mathrm{rank}_\varphi(x)$ such that $\forall x \, \mathrm{Ord}(\mathrm{rank}_\varphi(x))$ and $\varphi(x, y) \Rightarrow \mathrm{rank}_\varphi(x) < \mathrm{rank}_\varphi(y)$.

[Then most uses of recursion on $\varphi(x, y)$ can be replaced by recursion on the ordinals rank_φ, which may be easier to handle.]

(5) Cantor normal form for ordinals. Show that every ordinal α can be written in the form of a finite sum $\alpha = \omega^{\gamma_1} + \omega^{\gamma_2} + \ldots + \omega^{\gamma_j}$ with $\gamma_1 \geq \gamma_2 \geq \ldots \geq \gamma_j$, or (by collecting together equal terms) as $\alpha = \omega^{\beta_1}.n_1 + \omega^{\beta_2}.n_2 + \ldots + \omega^{\beta_k}.n_k$, where $\beta_1 > \beta_2 > \ldots > \beta_k$ and each $n_i \in \omega$ (and these are the operations of ordinal exponentiation etc.)

[Either of these forms is called the *Cantor normal form* for α. First show that if $\delta < \alpha$ then there is a unique ordinal $\beta \leq \alpha$ such that $\delta + \beta = \alpha$. Let γ_1 be $\sup\{\gamma \mid \omega^\gamma \leq \alpha\}$, and if $\omega^{\gamma_1} < \alpha$ let α_1 be such that $\omega^{\gamma_1} + \alpha_1 = \alpha$. Show that $\alpha_1 < \alpha$, since otherwise we would have $\omega^{\gamma_1+1} \leq \alpha$; and then repeat starting with α_1 instead of α. We shall get $\alpha_2 < \alpha_1 < \alpha$ etc., and this sequence must be finite since the ordinals are well-ordered. But the process can only stop when we get to 0.]

(6) Indecomposable ordinals. An ordinal α is called *additively indecomposable* if $\beta < \alpha \Rightarrow \beta + \alpha = \alpha$, (so that α cannot be written as the sum of two ordinals both less than α); it is *multiplicatively indecomposable* if $0 < \beta < \alpha \Rightarrow \beta.\alpha = \alpha$ (so that $\beta, \gamma < \alpha \Rightarrow \beta.\gamma < \alpha$).

Show that an ordinal $\alpha > 1$ is additively indecomposable if it is of the form $\alpha = \beta\omega$ for some β. Hence show that $\alpha > 1$ is additively indecomposable if and only if it is of the form ω^δ for some δ.

Show that an ordinal $\alpha > 2$ is multiplicatively indecomposable if it is of the form $\alpha = \beta^\omega$ for some β. Hence show that $\alpha > 2$ is multiplicatively indecomposable if and only if it is of the form ω^{ω^δ} for some δ.

[Use the Cantor normal form. [Sie58] gives more details.]

DEVELOPING MATHEMATICS WITHIN ZFC

6.7 Cardinals as initial ordinals

In 4.3 we introduced the naive notion of a cardinal; the usual definition for this naive notion uses the *initial* ordinals.

Definition 6.7.1 $\text{Init}(\alpha)$ for $\text{Ord}(\alpha) \wedge \forall \beta < \alpha (\beta \not\sim \alpha)$ (α is an *initial* ordinal).

The word *initial* is used here for the following reason: we can think of the ordinals as being partitioned into classes according to their cardinality. We have seen in 6.2.15(6) that the finite ordinals will each be in a class on its own; but $\omega, \omega+1, \omega+2, \ldots$ are all countable and will go in the same class (which Cantor called the *second number class*, finite ones forming the first). As we shall show, there must be many other classes for infinite ordinals. It is easy to see from the Schröder–Bernstein theorem 4.2.5 that if $\alpha < \gamma < \beta$ and $\alpha \sim \beta$ then also $\alpha \sim \gamma$, so each of these classes of ordinals must form an interval in the standard ordering of the ordinals. The first member of each such interval will be an initial ordinal, since it will not be of the same cardinal as any smaller ordinal.

Using the axiom of choice in 7.2 we shall see that every set is similar to some ordinal, and hence to a unique initial ordinal. So we can use these initial ordinals as representatives for each class of cardinals; we shall take $|x|$ to be the initial ordinal similar to x, for each set x.

It will follow from this that there must be many initial ordinals; but in fact we can prove already, without the axiom of choice, that there are many initial ordinals. A clear way to set this out is provided by Hartogs's aleph function (see [Har15]):

Definition 6.7.2 For any set x, $\aleph(x) = \{\alpha \mid \alpha \preccurlyeq x\}$.

Lemma 6.7.3 *For each set x, $\aleph(x)$ is an initial ordinal.*

Proof First we should show that $\aleph(x)$ is always a set. It is clearly a transitive collection of ordinals, from the definition, so it is either an ordinal or, if it were a proper class, it would be the collection of all ordinals. But for each $\alpha \in \aleph(x)$, suppose $g : \alpha \to x$ is an injection, and let $\mathbf{r}_g \subseteq (x \times x)$ be $\{\langle g(\beta), g(\gamma)\rangle \mid \beta < \gamma < \alpha\}$. Then $\mathbf{r}_g$ is a well-ordering of a subset of x in order-type α. Each such $\mathbf{r}_g$ corresponds to a unique α, and is a subset of the set $\mathcal{P}(x \times x)$, and so by the replacement axiom we shall get only a set of possible α's, as required.

Now if $\beta \sim \alpha$, and $\beta \in \aleph(x)$, then also $\alpha \in \aleph(x)$. So if $\aleph(x)$ is not an initial ordinal, we would have $\beta \in \aleph(x)$ with $\beta \sim \aleph(x)$ and hence $\aleph(x) \in \aleph(x)$, a contradiction. $\square$

Note that the same proof shows that we cannot have $\aleph(x) \sim x$. But if $x \sim \alpha$ then $\alpha \in \aleph(x)$, and so $\aleph(x)$ always gives us an initial ordinal which is the first ordinal which cannot be mapped into x. If x is similar to an initial ordinal (which it *always* is, if we assume the axiom of choice), then $\aleph(x)$ must be the next larger initial ordinal. We use this to list the infinite initial ordinals by transfinite induction:

Definition 6.7.4 (Cantor's alephs) The function $\aleph$ from ordinals to initial ordinals is given by:

$$\aleph_0 = \omega; \quad \aleph_{\alpha+1} = \aleph(\aleph_\alpha); \quad \text{and}$$

$$\aleph_\lambda = \bigcup_{\alpha<\lambda} \aleph_\alpha \quad \text{for limit } \lambda.$$

The proof that $\aleph_\lambda$ is an initial ordinal, for limit λ, is left as exercise 6.7.8(1).

This is Cantor's listing of the initial ordinals, or cardinals. An alternative notation is to write $\omega_\alpha = \aleph_\alpha$; originally the symbol ω_α was introduced to denote an ordinal (an initial ordinal, just as above), while $\aleph_\alpha$ denoted a cardinal, thought of in the naive manner introduced in 4.3. Nowadays the informal, naive use is usually abandoned in favour of the initial ordinals, and the two notations are used interchangeably.

The different uses of the symbol $\aleph$ should not confuse; really the use in 6.7.2 is only an intermediate, temporary use for us, as we are interested in set theory with the axiom of choice and, in that setting, the only use made of Hartogs's function is to define Cantor's alephs.

6.7.5 Cardinal arithmetic of the alephs

Under the axiom of choice, when all infinite cardinals can be taken as alephs, cardinal arithmetic becomes trivial for addition and multiplication. Both operations simply give the maximum:

Lemma 6.7.6 *For all α, the cartesian product $\aleph_\alpha \times \aleph_\alpha \sim \aleph_\alpha$. Hence for all α and β, cardinal addition and multiplication are given by $\aleph_\alpha + \aleph_\beta = \aleph_{\max(\alpha,\beta)}$ and $\aleph_\alpha \times \aleph_\beta = \aleph_{\max(\alpha,\beta)}$.*

Proof We need to define a bijection from the cartesian product $\aleph_\alpha \times \aleph_\alpha$ to $\aleph_\alpha$. We know that there is such a map for $\aleph_0$, and we shall proceed by induction on α; so we shall assume the result for all smaller ordinals.

We consider the following well-ordering of pairs of ordinals below $\aleph_\alpha$: for β, γ, β', $\gamma' < \aleph_\alpha$, define

$$\langle \beta, \gamma \rangle < \langle \beta', \gamma' \rangle \Leftrightarrow \max(\beta, \gamma) < \max(\beta', \gamma'); \quad \text{or}$$
$$\max(\beta, \gamma) = \max(\beta', \gamma') \text{ and } \beta < \beta'; \quad \text{or}$$
$$\max(\beta, \gamma) = \max(\beta', \gamma') \text{ and } \beta = \beta' \text{ and } \gamma < \gamma'.$$

DEVELOPING MATHEMATICS WITHIN ZFC

To see that this is a well-ordering, it is easiest to think in terms of descending sequences; an infinite descending sequence of pairs under $\prec$ would either involve infinitely many steps where the maximum decreased, or with fixed maximum, where the first elements decrease, or with one fixed first element, where the second elements decrease. Each way involves an infinite descending sequence of ordinals, which is impossible since the ordinals are well-ordered. (Of course, the axiom of choice is needed to get an infinite descending sequence in an arbitrary non-well-ordering; this proof can be set out more directly so that it clearly does not use the axiom of choice. See exercise 6.7.8(2).)

Now we apply 6.6.11 to get an isomorphism $g : \langle \aleph_\alpha \times \aleph_\alpha, \prec \rangle \to \langle \xi, < \rangle$ for some ordinal ξ. Since we can get from g an injection of $\aleph_\alpha$ into ξ we must have $\aleph_\alpha \leq \xi$; the result will follow when we show we cannot have $\aleph_\alpha < \xi$. We get a contradiction from supposing that $g(\langle \eta, \zeta \rangle) = \aleph_\alpha$ for some $\eta, \zeta < \aleph_\alpha$ (which must happen if $\aleph_\alpha$ belongs to the range of g). For, let $\delta = \max(\eta, \zeta)$ and let $y = \{\langle \eta', \zeta' \rangle \mid \langle \eta', \zeta' \rangle \prec \langle \eta, \zeta \rangle\}$. Then $y \subseteq (\delta+1) \times (\delta+1)$, and $g : y \to \aleph_\alpha$ onto, and $\delta + 1 < \aleph_\alpha$. So if β is the initial ordinal similar to $\delta + 1$, then $(\delta+1) \times (\delta+1) \sim \beta \times \beta$ and the induction hypothesis gives $\beta \times \beta \sim \beta$. Together this gives a map from β onto $\aleph_\alpha$, contradicting $\aleph_\alpha$ being an initial ordinal. □

6.7.7 Cardinal exponentiation

In contrast with cardinal addition and multiplication, cardinal exponentiation remains an interesting operation, and a large part of the problems of modern set theory can be presented in terms of trying to understand it. Some of the methods are introduced in chapter 8.

6.7.8 Exercises

(1) Show that $\aleph_\lambda$ is an initial ordinal, for λ a limit ordinal.
    [Such *limit cardinals* are usually singular; see 7.4.6(2), and also 9.4 for further properties.]
    Show further that there are many *fixed points* of the alephs, i.e. show that for any ordinal α there is $\lambda > \alpha$ such that $\lambda = \aleph_\lambda$.
    [Define $\alpha_0 = \alpha + 1$ and for $n < \omega$ let $\alpha_{n+1} = \aleph_{\alpha_n}$. Show that $\lambda = \bigcup_{n<\omega} \alpha_n$ satisfies $\lambda = \aleph_\lambda$.]
(2) Show that the ordering $\prec$ in 6.7.6 is a well-ordering of $\aleph_\alpha \times \aleph_\alpha$.
    [This is very similar to the proofs that the orderings given in 5.4.6 are well-orderings; but it is not in fact the same as any of the orderings used there.]
(3) Every construction of the reals makes use at some point of the power-set axiom, usually to justify use of the set of all functions from ω to the rationals, or directly to give the reals as a set of subsets of the rationals.

Show that this is an essential use; show that every set which can be shown to be a set without any use of the power-set axiom must be countable.

[The model used to show this is the model of *hereditarily countable sets*, which can be defined in a manner parallel to the cumulative types as follows:

Let $H_0 = V_\omega$, the collection of all hereditarily finite sets (as in 3.2.3(3) and 6.6.13(3)). Then for α a countable ordinal, define

$H_{\alpha+1}$ is the set of all countable subsets of H_α; and

$$H_\alpha = \bigcup_{\beta < \alpha} H_\beta \text{ for } \alpha \text{ a limit ordinal.}$$

Now if $H = \bigcup_{\alpha < \omega_1} H_\alpha$, H is the collection of all hereditarily countable sets.]

Show that H is a model of all the axioms of ZFC except the power-set axiom.

[Show that H is closed under countable subsets, and that every member of H is countable. Hence the replacement axiom will hold. But note that H itself is not countable; indeed the levels H_α will not be countable for $\alpha \geq 1$. Thus we will have $\mathcal{P}(\omega) \subset H_1$, but $\mathcal{P}(\omega) \notin H$; and further $H_1 = V_{\omega+1} \notin H$, but $H_2 \neq V_{\omega+2}$ since e.g. $H_1 \notin H_2$.]

Show that the structures ω, $\mathbb{Z}$ and $\mathbb{Q}$ will be in H; and that although each Dedekind section of $\mathbb{Q}$ and each Cauchy sequence of rationals will be in H, $\mathbb{R}$ itself cannot be represented in H.

[Simply because $\mathbb{R}$ is uncountable.]

7

The axiom of choice

It is important to realize when thinking about the axiom of choice that it can be proved using just first-order logic, together with the construction of pair-sets and finite unions; for *finite* sets, see exercise 7.1.7(1). Indeed, this is an important part of its justification, and any discussion of its justification which loses sight of this will probably be of little value. It follows that its application (if not redundant) will always be to infinite sets.

We start by looking at more technical questions in this chapter, and we first give some simple variants of the axiom. Then we shall prove what was historically its most striking consequence, Zermelo's well-ordering theorem. We also give some other, mathematically interesting variants, and some examples of their use.

7.1 Simple forms

7.1.1 AC1. The axiom of choice (AC) was stated in A10, 3.1.10 in the form

$$\forall x \forall y (x \in a \land y \in a \land x \neq y \Rightarrow x \cap y = \emptyset) \land \forall x (x \in a \Rightarrow x \neq \emptyset)$$
$$\Rightarrow \exists c \forall x [x \in a \Rightarrow \exists u (c \cap x = \{u\})].$$

Every disjointed set has a choice set. This is the simplest form to write in the primitive notation, without need for any definitions; but it is much more convenient in practice to use the form:

7.1.2 AC2. Choice functions:

$$\forall X \exists f (\text{Func}(f) \land \text{dom}(f) = X \land \forall x \in X (x \neq \emptyset \to f(x) \in x)).$$

Every set has a choice function, where a choice function takes as value a member of the argument (provided that argument is not empty), and its domain is the given set X. Note that some authors will make a different decision about how to treat the empty set, and will state the axiom in the form: every set of non-empty sets has a choice function.

Let us first prove the equivalence of these:

Lemma 7.1.3 *AC1 if and only if AC2.*

Proof Assuming AC1, and given X, first consider the case $\emptyset \notin X$ and let $X' = \{(\{x\} \times x) \mid x \in X\}$. Then this X' is disjointed: for if $u \in (\{x\} \times x)$, then u is of the form $\langle z, t \rangle$ with $z = x$ and $t \in x$. So $u \in (\{x\} \times x) \cap (\{y\} \times y) \Rightarrow u = \langle z, t \rangle$ with $z = x$ and $z = y$, i.e. $x = y$ so that also $(\{x\} \times x) = (\{y\} \times y)$. Since we are assuming $\emptyset \notin X$, also $\emptyset \notin X'$ and X' is disjointed. So we can apply AC1, and let C be a choice set for X'. Then C will have exactly one member from each $(\{x\} \times x) \in X'$; this member of C will be of the form $\langle x, t \rangle$ with $t \in x$. So C is a set of ordered pairs with just one pair for each first member, i.e. C is a function, and its domain will be the set X, and for each $x \in X$, $C(x) = t \in x$ if $\langle x, t \rangle \in C$. In other words, C will itself be a choice function for X.

To deal with the case that $\emptyset \in X$, we first take $X_0 = X - \{\emptyset\}$ and find a choice function C_0 for that; then e.g. $C = C_0 \cup \{\langle \emptyset, \emptyset \rangle\}$ will be a choice function for X (since no restriction is put on the value given by a choice function when it takes for its argument the empty set).

That completes AC1 $\Rightarrow$ AC2; the converse is simpler. Assuming AC2, and given a disjointed set A, suppose that f is a choice function for A. Then range f will be the required choice set for A. □

Another simple form of the axiom of choice was given by Russell, and he called it the multiplicative axiom (because of its use in cardinal multiplication). We use the notation for families here (see 10.2.4).

7.1.4 AC3. Multiplicative axiom:

$$\forall i \in I(X_i \neq \emptyset) \rightarrow \exists f(f \in \underset{i \in I}{\mathsf{X}} X_i).$$

Or to take the contrapositive, the cartesian product of a family $\langle X_i \mid i \in I \rangle$ is empty only if one of the factors, X_i for some $i \in I$, is empty.

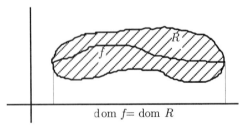

Diagram 7.1.5

7.1.5 AC4. Uniformization. Another simple form can be stated as:

$$\text{Rel } R \rightarrow \exists f \subseteq R(\text{dom } f = \text{dom } R \wedge \text{Func } f).$$

THE AXIOM OF CHOICE   111

Every relation can be *uniformized* by a function, i.e. a function which is a subset of the given relation with the same domain. This can be thought of in graphical terms, see diagram 7.1.5

Lemma 7.1.6 *AC3 and AC4 are equivalent to the axiom of choice.*

Proof We actually prove AC2 ⇒ AC3 ⇒ AC4 ⇒ AC2.

AC2 ⇒ AC3: Given the indexed family $\langle X_i \mid i \in I \rangle$, let Y be the set of its members, i.e. $Y = \{X_i \mid i \in I\}$, and let f be a choice function for Y (by AC2). Then g given by $g(i) = f(X_i)$ will satisfy $g \in \bigtimes_{i \in I} X_i$, unless $X_i = \emptyset$ for some $i \in I$.

AC3 ⇒ AC4: Given a relation R, let $I = \text{dom}\, R$ and set $X_i = \{x \mid \langle i, x \rangle \in R\}$ for each $i \in I$. Then $X_i \neq \emptyset$ for $i \in I$ by the definition of $\text{dom}\, R$, so (by AC3) let $g \in \bigtimes_{i \in I} X_i$. Then $g \subseteq R$ and g is a function with domain $I = \text{dom}\, R$, i.e. g uniformizes R.

AC4 ⇒ AC2: Given X, define $R = \{\langle x, y \rangle \mid y \in x \in X\}$, or in other terms $R = \bigcup \{ \{x\} \times x \mid x \in X \}$. Then R is a relation, and $\text{dom}\, R = X - \{\emptyset\}$. So if f uniformizes R, f will be a choice function for $X - \{\emptyset\}$; if $\emptyset \in X$ we can add $\langle \emptyset, \emptyset \rangle$ to f as we did in 7.1.3 to get a choice function for X. □

Direct proofs of AC1 ⇒ AC3, AC1 ⇒ AC4, AC3 ⇒ AC2, and AC4 ⇒ AC3 can all be given; see 7.1.7(2). But it seems very hard to give simple direct proofs of AC3 ⇒ AC1 and AC4 ⇒ AC1 without in fact going via AC2, i.e. without in effect first giving a choice *function*, whose range will be the desired choice *set*. (But see 7.3.3 below, or 7.3.4(3) (from Tukey–Teichmüller form) for a direct proof of AC1 from another form.)

7.1.7 Exercises

(1) Prove the axiom of choice for finite sets in the form of AC1 or AC2 using only the pair-set and sum-set axioms.

[For AC1, first prove by induction on n that, given $\exists y \in x_i$ for $i = 1, \ldots, n$, then $\exists c \exists y_1 \ldots \exists y_n (c = \{y_1, \ldots, y_n\} \land y_1 \in x_1 \land \ldots \land y_n \in x_n)$. Hence if $a = \{x_1, \ldots, x_n\}$ and is disjointed, a choice set exists. A similar proof can be given for AC2.

These steps can then be put more formally to prove by induction on ω (and hence using the axiom of infinity), that $\forall n \in \omega (a \sim n \Rightarrow \text{AC1})$ and $\forall n \in \omega (X \sim n \Rightarrow \text{AC2})$.]

(2) Give direct proofs of AC1 ⇒ AC3, AC1 ⇒ AC4, AC3 ⇒ AC2, and AC4 ⇒ AC3.

[Given the indexed family $\langle X_i \mid i \in I \rangle$ with $X_i \neq \emptyset$ for $i \in I$, form the set $\{(\{i\} \times X_i) \mid i \in I\}$ and apply AC1 to it, since it is disjointed, or apply AC4 to it, since it is a relation, in each case to get AC3. Given a relation R, let $R_x = \{y \mid \langle x, y \rangle \in R\}$ for $x \in \text{dom}\, R$, and apply AC1 to

$\{(\{x\} \times R_x) \mid x \in \operatorname{dom} R\}$ to get AC4. Given a set X with $\emptyset \notin X$ take $\{\langle x, x \rangle \mid x \in X\}$ as an indexed family and apply AC3 to get AC2.]

7.2 The well-ordering theorem

We gave the definition of well-orderings in 5.4, and noted in 6.6.10 that set X has a well-ordering if and only if $X \sim \alpha$ for some ordinal α. Here we shall prove, from the axiom of choice, that every set is similar to an ordinal. This is a strong form of the well-ordering theorem; the original theorem proved by Zermelo was just that every set is well-ordered by some relation, and this is a weaker theorem. In fact our proof makes essential use of the replacement axiom (in making use of transfinite induction, theorem 6.6.4), whereas Zermelo's original proof did not need the replacement axiom.

Theorem 7.2.1 *(AC) Every set is similar to some ordinal.*

As is customary in much of mathematics, and as we shall do (at least in important cases) for the remainder of this book, we note where our results rely on the axiom of choice by the note (AC) before any formal statement which makes use of any form of the axiom.

Proof We show AC2 $\Rightarrow \forall X \exists \alpha (X \sim \alpha)$.

Given X, take f to be a choice function on the set $\mathcal{P}(X)$, so if $Y \subseteq X$, $Y \neq \emptyset$, then $f(Y) \in Y$. Now we define a function h by transfinite induction on the ordinals by:

$$h(\beta) = f(X - h''\beta) \quad \text{until} \quad h''\beta = X.$$

Let α be the first ordinal such that $h''\alpha = X$. There must be such an α, since otherwise h is one–one from On, the class of all ordinals, into X, contradicting Burali-Forti's paradox. (This is just the same argument as in the proof of 6.6.11.)

Then $h : \alpha \to X$ will be one-one onto X, since if $\gamma < \beta < \alpha$ then $h(\beta) = f(X - h''\beta) \in X - h''\beta$; but $h(\gamma) \in h''\beta$, so $h(\beta) \neq h(\gamma)$. □

We prove the converse of this from the weaker statement, since that is then a stronger result:

Theorem 7.2.2 $\forall X \exists r (r \text{ wo } X) \Rightarrow AC2$ *(where we write r wo X for: $\langle X, r \rangle$ is well-ordered).*

Proof Given X, let $Y = \bigcup X$ and assume r wo Y. Then we use r to produce a choice function for X by taking the r-least member of each $x \in X$: if $x \neq \emptyset$

THE AXIOM OF CHOICE

then we define $f(x) = u$ where $u \in x$ is the r-least; since $x \subseteq Y$ this will exist and be unique unless $x = \emptyset$. A formal definition of f would be:

$$f = \{\langle x, u\rangle \mid x \in X \wedge u \in x \wedge \forall v \in x \neg(vru)\}.$$

If $\emptyset \in X$ we add $\langle \emptyset, \emptyset \rangle$ to f to complete the choice function as usual. □

7.2.3 Comparability of cardinals

Given the well-ordering theorem, we can now formalize the work in 4.3 which assumed a definition of *cardinal of a set* with appropriate properties; we can now use the initial ordinals as cardinals, and prove the required results (as noted in 6.7.1). An important corollary is:

Corollary 7.2.4 *(AC) Any two sets are comparable in the sense of cardinality; i.e. for any two sets X and Y, $X \preccurlyeq Y$ or $Y \preccurlyeq X$.*

Proof Immediate from 7.2.1 and 6.5.3 (the comparability of ordinals). □

The converse of 7.2.4, that the axiom of choice follows from the comparability of cardinals, is the next exercise:

7.2.5 Exercises

(1) Show that x can be well-ordered if there is an injection $f : x \to \alpha$ for some ordinal α; or if there is any function $g : \beta \to y$ with β an ordinal and $x \subseteq \text{range}(g)$.

[In both cases the well-ordering transfers from the ordinal to x, in the first case using $z \preccurlyeq z' \Leftrightarrow f(z) < f(z')$ (for $z, z' \in x$), and in the second case by defining $f(z) = \delta$ if $z \in x$ and $\delta \in \beta$ is the least for which $g(\delta) = z$, so that $f : x \to \beta$ is an injection.]

Hence show that the axiom of choice follows from the assumption that any two sets are comparable in the sense of cardinality.

[Compare x with $\aleph(x)$ as defined in 6.7.2. It follows from what was done there that we cannot have $\aleph(x) \preccurlyeq x$, hence from comparability of cardinals we get $x \prec \aleph(x)$; but $\aleph(x)$ is an ordinal.]

(2) Zermelo's proof of the well-ordering theorem. Prove that AC2 $\Rightarrow$ $\forall X \exists r (r \text{ wo } X)$ without using the replacement axiom.

[This is very much a matter of setting out the proof of transfinite induction just as far as is needed to get a well-ordering of X. The good parts, which will be put together to get the well-ordering, will be well-orderings of subsets of X which have the property that the next element is always that one which the given choice function on $\mathcal{P}(X)$ chooses from the remaining elements of X (i.e. those which have not yet appeared in the well-ordering). These good parts will always be subsets of $X \times X$ and so the collection of all of them will be a set, without any need for replacement.]

7.3 Maximal principles and Zorn's lemma

Several important forms of the axiom of choice, of which the first was given by Hausdorff ([Hau35]), involve the existence of maximal objects of various kinds. The relation with respect to which the objects are maximal may just be the subset relation, as in Hausdorff's form, or it may be a given partial ordering, as in Zorn's lemma. These maximal objects are usually not unique. It is quite clear that not every partial ordering need have maximal elements; ω is the simplest counterexample. Hence the importance of the hypothesis of Zorn's lemma.

Theorem 7.3.1 *(AC) (Hausdorff) Every partial ordering has a maximal chain.*

Here a *chain* in a partial ordering $\langle X, < \rangle$ is a subset $C \subseteq X$ which is totally ordered by $<$ (see 5.1.5). A singleton will always count as a chain, so there must be many chains; and *maximal* here means maximal under the subset relation, i.e. a chain which is not a proper subset of any other chain. Unless the given partial ordering is total (i.e. already a chain), the maximal chain will not be unique (exercise 7.3.4(2)).

Theorem 7.3.2 *(AC) (Zorn). If $\langle X, < \rangle$ is a partial ordering in which every chain has an upper bound, then X has a maximal element (under $<$).*

It is immediate that 7.3.1 (Hausdorff's principle) implies 7.3.2 (Zorn's lemma) since an upper bound of a maximal chain must be a maximal element (and the maximum member of that chain), otherwise adding the upper bound would extend the chain. The converse we leave as exercise 7.3.4(1).

Proof of 7.3.2 Given $\langle X, < \rangle$ a partial ordering satisfying the hypothesis of Zorn's lemma, we take f a choice function on $\mathcal{P}(X)$ using AC2. Given any subset Y of X, define X above Y by

$$X \text{ above } Y = \{x \in X \mid \forall y \in Y (y < x \land y \neq x)\},$$

i.e. the set of all strict upper bounds of Y in X. Then define h by transfinite induction on the ordinals by:

$$h(\beta) = f(X \text{ above } h''\beta) \quad \text{until} \quad X \text{ above } h''\beta = \emptyset.$$

THE AXIOM OF CHOICE

Let α be the first ordinal where X above $h''\alpha = \emptyset$; there must be such an α since h is one–one below α and so would give a one–one map of all ordinals into X if we never found such an α (i.e. we use again the same argument we used in 7.2.1 and 6.6.11).

Now $h''\alpha$ is a chain in X by its construction (since at each step $h(\beta)$ is an upper bound of $h''\beta$). So we can apply the hypothesis that every every chain in $\langle X, < \rangle$ has an upper bound, and assume that x is an upper bound of $h''\alpha$. x cannot be a strict upper bound of $h''\alpha$ since there are none (since X above $h''\alpha$ is empty); so it must be the greatest member of $h''\alpha$. If it were not maximal in $\langle X, < \rangle$, say if $y \in X$ had $x < y$, then we would have $y \in X$ above $h''\alpha$; contradiction. So this x must be maximal, as required. □

To complete the proof that Zorn's lemma and Hausdorff's maximal principle are equivalent to the axiom of choice, and not just consequences, we prove:

Theorem 7.3.3 *Zorn's lemma implies AC1.*

Proof Given a disjointed set A, let

$$X = \{c \subseteq \bigcup A \mid \forall a \in A (\text{Card}(c \cap a) \leq 1)\},$$

partially ordered by $\subseteq$. The members of X are partial choice sets on A, and it is clear that there must be many such, e.g. finite ones. It is also clear that if c is maximal in X, then c must in fact be a choice set for the whole of A, since if $a \in A$ were such that $\text{Card}(c \cap a) = 0$ we could extend c by adding a member of a. (We know $a \neq \emptyset$; let $y \in a$ and let $c' = c \cup \{y\}$, then $c' \in X$ and $c \subset c'$.)

It remains to show that X satisfies the hypothesis of Zorn's lemma, under $\subseteq$; i.e. that any chain in X has an upper bound in X. This is probably the commonest, and certainly the simplest case where we need this property. We suppose that $\mathcal{C} \subseteq X$ is a chain under $\subseteq$, and show that $\bigcup \mathcal{C}$ is the required upper bound. In fact the definition of *upper bound* can be thought of as a generalization of the definition of $\bigcup$; it is immediate that $c \in \mathcal{C} \Rightarrow c \subseteq \bigcup \mathcal{C}$ and that if $c \in \mathcal{C} \Rightarrow c \subseteq d$ then $\bigcup \mathcal{C} \subseteq d$. But we do have more to prove, since we must show also that $\bigcup \mathcal{C}$ is in fact a *member* of X. So suppose $a \in A$; we must show that $\text{Card}(\bigcup \mathcal{C} \cap a) \leq 1$. Suppose that both u and v are in $\bigcup \mathcal{C} \cap a$. Then $u \in c \in \mathcal{C}$ and $v \in c' \in \mathcal{C}$, for some c, c'. At this point we use the fact that $\mathcal{C}$ is a *chain* (it is not used anywhere else, so we'd better use it somewhere!). We must have either $c \subseteq c'$ or $c' \subseteq c$ since $\mathcal{C}$ is a chain under $\subseteq$; assume $c \subseteq c'$. Then we must have both $u \in c' \cap a$ and $v \in c' \cap a$; but $c' \in X$ and so $\text{Card}(c' \cap a) \leq 1$. Hence $u = v$ and we have shown that $\text{Card}(\bigcup \mathcal{C} \cap a) \leq 1$ also. So $\bigcup \mathcal{C} \in X$ and we are done. □

7.3.4 Exercises

(1) Prove directly that Zorn's lemma implies Hausdorff's principle.

[Given a partial ordering $\langle X, < \rangle$, consider the partial ordering of all chains in $\langle X, < \rangle$ ordered by the subset relation. Show that the union of any chain of chains is again a chain, and hence Zorn's lemma applies.]

Prove also directly that Zorn's lemma implies the well-ordering theorem.

[Consider the collection of all well-orderings of subsets of X (where X is the set to be well-ordered). There are clearly many such well-ordered subsets, e.g. all finite ones. Now the important point is to consider this collection as partially ordered by *initial segments*, i.e. we take $\mathbf{r} \preccurlyeq \mathbf{s}$, where $\mathbf{r}$ and $\mathbf{s}$ are well-orderings of subsets of X, to mean that $\mathbf{r}$ is an initial segment of $\mathbf{s}$ (in particular, if $x \in \text{dom}\,\mathbf{r}$ and $y\,\mathbf{s}\,x$ then $y\,\mathbf{r}\,x$ also). Show that the union of a chain under this relation will again be a well-ordering of a subset of X.

(Note that this would fail if we took the same collection ordered by the subset relation; it is easy to construct a chain of well-orderings under inclusion, such that the union will have a descending chain. We can form such a chain by simply adding new members at the beginning each time.)

The final step is to show that a maximal element in this collection must be a well-ordering of the whole of X, since, if not, we can extend it by adding some new element at the end.]

(2) Prove that a partial ordering has a unique maximal chain if and only if the partial ordering is linear. Where is the axiom of choice needed?

[Show that any chain can be extended to a maximal chain.]

(3) The Tukey–Teichmüller lemma. Set A is *of finite character* means: $X \in A$ if and only if every finite subset of X is in A. Show, using the axiom of choice, that any set of finite character must have a maximal element (under $\subseteq$).

[Take the union of a maximal chain in A.]

Show that AC1 follows from the Tukey–Teichmüller lemma.

[Use the same proof as in 7.3.3; show that the set X used there has finite character (so that the Tukey–Teichmüller lemma can be used in place of Zorn's lemma to get a maximal element).]

(4) Show that the hypothesis of Zorn's lemma is not *necessary* for a partial ordering to have a maximal element.

[An isolated element in a partial ordering, not related to any others, will automatically count as maximal; so the simplest partial ordering with a maximal element and *not* satisfying the hypothesis of Zorn's lemma is probably the ordering ω together with one further isolated element.]

(5) Prove AC2 directly from Hausdorff's maximal principle.

[If X is the set for which a choice function is required, take the partial

THE AXIOM OF CHOICE   117

ordering of all functions which are choice functions for a subset of X, ordered by inclusion. Show that the union of a maximal chain must be a choice function for the whole of X.]
(6) Show the following is equivalent to the axiom of choice: every partially ordered set $\langle X, < \rangle$ has a cofinal well-ordering.

[This is part of the proof given for Zorn's lemma in 7.3.2, and it is clear that it embodies sufficient of the axiom of choice to prove Zorn's lemma. It is proved there by transfinite induction from AC2. Another proof is from Zorn's lemma: take the collection of all subsets of X which are well-ordered by $<$; this collection partially ordered by initial segments, as in (1) above, so that it satisfies the hypothesis of Zorn's lemma. Then show that a maximal element must be cofinal (otherwise it could be extended).]

7.4 Simple consequences of the axiom of choice

So far we have only presented equivalents of AC1 (equivalent given the rest of our axioms; it is clear that the reliance on other axioms varies greatly). We now present a few of the many results known which need the axiom of choice for their proof, but from which we cannot derive the full axiom of choice. These could be regarded as partial choice principles but some are more deserving of that name than others. More such are in the exercises. First two uses that were noted in chapter 4:

7.4.1 The smallest infinite set

We commented in 4.1.5 that the axiom of choice is needed in order to prove that ω is the smallest infinite set. This is now a simple corollary of the well-ordering theorem: if $X \sim \alpha$, then either $\alpha \in \omega$, in which case X is finite; or $\omega \leq \alpha$ and hence $\omega \preccurlyeq X$. So every set is either finite, or contains a denumerable set. (Note exercise 7.4.6(1) for a direct proof from AC1.)

7.4.2 Countable union of countable sets

Let $(A_n)_{n \in \omega}$ be a countable family of countable sets, and for each $n \in \omega$ let $X_n = \{g \mid g : \omega \to A_n \text{ one-one onto}\}$. Then $X_n \neq \emptyset$ for each $n \in \omega$, and we can apply AC3 to $\bigtimes_{n \in \omega} X_n$ to get f such that for each $n \in \omega$, $f(n) \in X_n$, i.e. $f(n) : \omega \to A_n$ as was required in 4.1.12.

An important use of this is in proving the regularity of aleph one, see exercise 7.4.6(2).

7.4.3 Dependent choices, DC

This is the following principle: suppose $A \neq \emptyset$ and $\mathbf{R}$ is some relation on A. Then:

$$\forall x \in A \exists y \in A(x \,\mathbf{R}\, y) \Rightarrow \exists f : \omega \to A[\forall n \in \omega(f(n) \,\mathbf{R}\, f(n+1))].$$

Here the sequence $\langle f(n) \mid n \in \omega \rangle$ is being chosen so that each choice of $f(n+1)$ depends on the previous choice $f(n)$ (and the relation $\mathbf{R}$ says how it is to depend). It is instructive to compare this with the recursion theorem 6.2.3: if $\mathbf{R}$ specifies exactly which member of A is to follow (so that $\mathbf{R}$ is in fact a function on A), then this statement is just a restatement of the recursion theorem. But if $\mathbf{R}$ allows for some choice, i.e. if $|\{y \mid x \,\mathbf{R}\, y\}| > 1$ for infinitely many $x \in A$, then the axiom of choice may be needed. The proof is indeed just a combination of the axiom of choice and the recursion theorem:

Proof Given relation $\mathbf{R}$ on A, let g uniformize $\mathbf{R}$ by AC4. Then the hypothesis $\forall x \in A \exists y \in A(x \,\mathbf{R}\, y)$ implies that $\operatorname{dom} \mathbf{R} = A$ and hence $g : A \to A$. So we can define $f : \omega \to A$ by:

$$f(0) = a_0,$$
$$f(n+1) = g(f(n)),$$

using the recursion theorem, with a_0 any member of A. Since $g \subseteq \mathbf{R}$, we get $f(n) \,\mathbf{R}\, f(n+1)$ as required. □

7.4.4 Descending chains

We noted in 5.4.2 that any linear ordering which is not a well-ordering must contain a descending chain. This is a corollary of dependent choices. Suppose that $\langle X, < \rangle$ is a linear order, and that $Y \subseteq X$ has no least member under $<$, i.e. $\forall x \in Y \exists z \in Y (z < x)$. Then take $\mathbf{R}$ as the strict converse of $<$, i.e. define $x \,\mathbf{R}\, z$ if and only if $z < x \land z \neq x$, for $x, z \in Y$. Now apply 7.4.3 and the resulting $f : \omega \to Y$ will be a descending chain.

7.4.5 Ultrafilters and prime ideals

Zorn's lemma can be used to prove the existence of maximal proper filters and ideals; it is then a matter of the appropriate lattice theory to complete the work left in 5.3.15. We give the basic ideas here for ideals, leaving the dual results for filters as exercises.

The first fact needed is: if $\mathcal{C}$ is a chain of ideals (in any lattice), then $\bigcup \mathcal{C}$ is also an ideal. (To show that if $x, y \in \bigcup \mathcal{C}$ then $x \lor y \in \bigcup \mathcal{C}$, we must make use of the fact that $\mathcal{C}$ is a *chain*, very much as it was used in the proof of 7.3.3.)

THE AXIOM OF CHOICE   119

This will then give the hypothesis of Zorn's lemma for the collections we shall need to consider, which will be collections of ideals ordered by inclusion.

Now in 5.3.15 we are given a distributive lattice $\langle A, \preccurlyeq \rangle$ and elements $a, b \in A$ with $a \preccurlyeq b, a \neq b$, and we have to show there is a prime ideal I in A with $a \in I$ and $b \notin I$. We take $\mathcal{J} = \{I \subseteq A \mid I \text{ is an ideal with } a \in I, b \notin I\}$. Then the principal ideal generated by a (i.e. $\{x \in A \mid x \preccurlyeq a\}$), will be in $\mathcal{J}$ (so $\mathcal{J}$ is not empty), and (using our first fact) we apply Zorn's lemma to $\mathcal{J}$ to get a maximal element J of $\mathcal{J}$ under inclusion. Then we must show that J is prime. Two steps give this:

(i) If $c \notin J$ then for some $y \in J$, $b \preccurlyeq c \vee y$.

For we know that the ideal $J + c = \{x \in A \mid x \preccurlyeq c \vee y \text{ for some } y \in J\}$ is not in $\mathcal{J}$, since J is maximal in $\mathcal{J}$. So we must have $b \in J + c$. (See 5.3.16(5) for the properties of $J + c$.)

(ii) Now if J is not prime, say $x \wedge z \in J$, but $x \notin J$ and $z \notin J$, we get a contradiction: by (i) we get $y, y' \in J$ with $b \preccurlyeq x \vee y$ and $b \preccurlyeq z \vee y'$. Hence $b \preccurlyeq (x \vee y) \wedge (z \vee y')$. But

$$(x \vee y) \wedge (z \vee y') \preccurlyeq (x \vee (y \vee y')) \wedge (z \vee (y \vee y'))$$
$$= (x \wedge z) \vee (y \vee y') \quad \text{by distributivity}$$
$$\in J \quad \text{since } J \text{ is an ideal.}$$

And this gives $b \in J$, which is the contradiction (since $J \in \mathcal{J}$). □

7.4.6 Exercises

(1) Use AC1 directly to prove that if X is not finite, there is an embedding of ω into X.

[Since X is not finite, for each $n \in \omega$ the set $X_n = \{f \mid f : n \to X\}$ is not empty. Let $A = \{X_n \mid n \in \omega\}$ and apply AC1; then to construct the required embedding out of a choice set for A is relatively straightforward.]

(2) Cofinality, and the regularity of successor cardinals. For any ordinal α, define the *cofinality* of α, $\operatorname{cf}(\alpha)$ as the smallest ordinal δ for which there is a function $f : \delta \to \alpha$ with range *cofinal* in α, i.e. for which $\beta < \alpha \Rightarrow \exists \gamma < \delta (\beta \leq f(\gamma))$.

Show that if α is a limit ordinal, then $\operatorname{cf}(\alpha)$ is an initial ordinal, i.e. a cardinal, and that $\operatorname{cf}(\operatorname{cf}(\alpha)) = \operatorname{cf}(\alpha)$.

[For successor α, $\operatorname{cf}(\alpha) = 1$ and is of little interest. But for any limit ordinal, its cofinality is of great interest.]

An ordinal α is called *regular* if $\operatorname{cf}(\alpha) = \alpha$ (and then it must be a cardinal); otherwise it is *singular*. Show that any successor cardinal $\aleph_{\alpha+1}$ must be regular.

[If not, its cofinality δ must be $\leq \aleph_\alpha$, and so we would have $f : \delta \to \aleph_{\alpha+1}$ with cofinal range, i.e. such that $\bigcup_{\gamma < \delta} f(\gamma) = \aleph_{\alpha+1}$. But each $|f(\gamma)| \leq \aleph_\alpha$, so using the axiom of choice as in 7.4.2 to choose a well-ordering of each $f(\gamma)$ of order-type $\leq \aleph_\alpha$ and using 6.7.6, we get $\left|\bigcup_{\gamma<\delta} f(\gamma)\right| = \aleph_\alpha$, contradicting the definition of $\aleph_{\alpha+1}$.]

Show that if λ is a limit ordinal, then $\mathrm{cf}(\aleph_\lambda) = \mathrm{cf}(\lambda)$. Hence limit cardinals will be singular, unless $\aleph_\lambda = \lambda$; but show that the cardinals constructed in 6.7.8(1) with this property are still singular, with cofinality ω.

[The construction in 6.7.8(1) can easily be extended to give limit cardinals with any pre-determined desired cofinality; but there is no way to construct *regular* limit cardinals. See 9.4.]

(3) Maximal ideals and filters. These were defined in 5.3.16(6).

Show that any lattice with $\mathbb{I}$ must have a maximal ideal; and a lattice with $\mathbb{O}$ must have a maximal filter.

[This is a direct application of Zorn's lemma. Take the collection of all proper ideals, ordered by inclusion; the proof that this satisfies the hypothesis of Zorn's lemma is set out in 7.4.5 above, except that we must use the existence of $\mathbb{I}$ to show that the union of a chain of proper ideals is still proper (since in a lattice with $\mathbb{I}$ an ideal is proper if and only if $\mathbb{I}$ is not in it). A maximal ideal is just a maximal element of the collection considered, by definition.]

(4) Construction of maximal ideals and filters by transfinite recursion.

The proof using Zorn's lemma that maximal ideals exist, in any lattice with $\mathbb{I}$, gives little control over the resulting ideal. More can be deduced if the collection considered is restricted: e.g. if we consider only proper ideals extending a given proper ideal, we prove that every proper ideal can be extended to a maximal ideal. But for more detailed control, we can use transfinite recursion.

Suppose that the elements of the lattice A are enumerated as $A = \{a_\alpha \mid \alpha < \kappa\}$ where $\kappa = |A|$. Taking I_0 as some starting ideal (or $I_0 = \emptyset$), define

$$I_{\alpha+1} = \begin{cases} I_\alpha + a_\alpha & \text{if } I_\alpha + a_\alpha \text{ is proper;} \\ I_\alpha & \text{otherwise;} \end{cases}$$

and $I_\lambda = \bigcup_{\alpha<\lambda} I_\alpha$ for limit λ.

Show that I_κ is a maximal ideal.

[Varying the order of enumeration of A will give variants of the maximal ideal; but more powerful variants come from varying the conditions used at each step.]

(5) König's lemma. This is a generalization of Cantor's theorem. For many years it was the only extra fact known about cardinal exponentiation, beyond the monotonicity properties of 4.3.6.

Let $(m_i)_{i \in I}$, $(n_i)_{i \in I}$ be two families of cardinals, with $m_i < n_i$ for each $i \in I$. Show (using the axiom of choice) that $\sum_{i \in I} m_i < \prod_{i \in I} n_i$.

[Here it is the strict inequality which is of interest. First, the axiom of choice is needed to choose two families of sets, $(x_i)_{i \in I}$ and $(y_i)_{i \in I}$, with the (x_i) disjoint and $|x_i| = m_i$, $|y_i| = n_i$ for $i \in I$. Given an embedding from x_i into y_i for each i, it is easy to construct an embedding from $\bigcup_{i \in I} x_i$ into $\bigtimes_{i \in I} y_i$. Show that any such embedding cannot be onto. Suppose given $f : \bigcup_{i \in I} x_i \to \bigtimes_{i \in I} y_i$; let $z_i = \{f(t)_i \mid t \in x_i\}$ and $u_i = y_i \setminus z_i$ for each $i \in I$. The hypothesis $m_i < n_i$ implies that $u_i \neq \emptyset$ for each $i \in I$, so apply AC3 to get $g \in \bigtimes_{i \in I} u_i$. Show that g is not in the range of f; for any $i \in I$, $f(i)_i \in z_i$ while $g_i \in u_i$, so $g \neq f(i)$.]

Hence show that if $m \geq 2$ and n is infinite, then $\mathrm{cf}(m^n) > n$.

[Take any $h : n \to m^n$ and show it cannot be cofinal; show that $\bigcup_{\alpha \in n} h(\alpha) < m^n$. Note that $|h(\alpha)| < m^n$ for $\alpha \in n$ and apply the result above with $m_\alpha = |h(\alpha)|$ and $n_\alpha = m^n$ for $\alpha \in n$; we get

$$\left| \bigcup_{\alpha \in n} h(\alpha) \right| \leq \sum_{\alpha \in n} |h(\alpha)| < \prod_{\alpha \in n} m^n = (m^n)^n = m^{n \cdot n} = m^n.$$

Note that Cantor's theorem follows immediately, taking $m = 2$.]

(6) Use Zorn's lemma to show that every partial ordering can be extended to a linear ordering; i.e. if $\langle P, \preccurlyeq \rangle$ is a partial ordering, there is a total ordering $\langle P, < \rangle$ with $\preccurlyeq \; \subseteq \; <$.

[One way is to apply Zorn's lemma to the collection (ordered by inclusion) of all partial orderings of P which extend $\preccurlyeq$. Show that this collection satisfies the hypothesis of Zorn's lemma (unions of chains are still partial orderings of P which extend $\preccurlyeq$), and that a maximal element must be total—for this, show that if a partial ordering is not total, say a and b are not related by **r**, then we can extend **r** to **r**′ by adding a **r**′ b, and making **r**′ transitive by also adding t **r**′ s for any $t, s \in P$ for which t **r** a and b **r** s.

Another way is to apply Zorn's lemma to the collection of all total orderings of a subset of P which extend $\preccurlyeq$ where they are defined. This requires a little more work to show that a maximal element must linearly order the whole of P.]

(7) König's infinity lemma. Show that if every chain and every antichain of a partially ordered set are finite, then the set is finite. (An *antichain* is a subset of any partial ordering which is completely unordered by the ordering.)

[The contrapositive really is König's infinity lemma (due to D. König, not to be confused with J. König in (5) above). To prove it, assume that $\langle P, \preccurlyeq \rangle$ is a partial ordering for which every antichain is finite, but with P infinite. Note that if $X \subseteq P$ then we can choose a maximal antichain in X, which will be a finite set $\{a_1, \ldots, a_k\}$ such that for any $x \in X$, either $x \preccurlyeq a_i$ or $a_i \preccurlyeq x$ for some $i \leq k$. To choose such a finite maximal antichain will not in fact need the axiom of choice—it will follow from the finite form of the axiom of choice as in 7.1.7(1); but in general it is an easy application of Zorn's lemma or of Hausdorff's maximal principle to show that any partially ordered set must have a maximal antichain.

So if we write $x \prec y$ for $x \preccurlyeq y \wedge x \neq y$, and for $X \subseteq P$ and $a_i \in P$ let

$$A^X(a_i) = \{y \in X \mid a_i \prec y\}, \quad B^X(a_i) = \{y \in X \mid y \prec a_i\},$$

(A for above, B for below), then we have

$$X = \bigcup_{i \leq k}(A^X(a_i) \cup B^X(a_i)) \cup \{a_1, \ldots, a_k\}$$

where $\{a_1, \ldots, a_k\}$ is the maximal antichain chosen in X. So if X is infinite, at least one of $A^X(a_i)$ and $B^X(a_i)$ is infinite for some $i \leq k$. (Note the use of 6.2.15(8), which showed that a finite union of finite sets is finite.)

Now we proceed to prove König's lemma as follows: start with $X_0 = P$ and define x_n and X_n recursively so that $x_n \in X_n$ is a point for which either $A^{X_n}(x_n)$ or $B^{X_n}(x_n)$ is infinite, and $X_{n+1} = A^{X_n}(x_n)$, or $B^{X_n}(x_n)$ if $A^{X_n}(x_n)$ is not infinite. Then $\{x_n\}$ will be an infinite chain in $\langle P, \preccurlyeq \rangle$.]

(8) *Closed unbounded sets and stationary sets.* Given a cardinal κ of uncountable cofinality, we define *closed unbounded subsets* of κ (usually abbreviated to *club* or sometimes *cub*) subsets) to be subsets $X \subseteq \kappa$ which are unbounded in κ, i.e. $\delta < \kappa \Rightarrow \exists \gamma \in X (\delta < \gamma)$, and *closed*, which we can define directly by: if $Y \subseteq X$, $Y \neq \emptyset$, and $\bigcup Y < \kappa$ then $\bigcup Y \in X$. (This coincides with being closed in the *order topology* on κ; another way to say it is that X is closed under taking limits, provided those limits are $< \kappa$, since the $\bigcup$ operation is the appropriate limit operation—see 6.5.8(1).)

Show that a set is club if and only if it is the range of some unbounded, increasing, continuous function $f : \lambda \to \kappa$, for some $\lambda < \kappa$ (where *continuous* means $f(\alpha) = \bigcup_{\delta < \alpha} f(\delta)$ for each limit $\alpha < \lambda$). Show that the intersection of two club subsets is again a club subset; and further that the intersection of any collection of $< \operatorname{cf} \kappa$ club subsets is again club.

[To show that the intersection of any collection of closed sets is closed is a standard exercise in topology, and is straightforward here. To show that it is unbounded, pick an ascending sequence of members of each of the sets in turn, repeating ω times; the limit will still be $< \kappa$ by the definition of cofinality, and will be in all of the sets by closure. Note how this can fail for κ of countable cofinality.]

In view of this, the collection of club subsets of κ generates a filter, the *club filter* on κ, which will be $< \operatorname{cf} \kappa$-closed. Sets which meet every member of this filter (equivalently, every club set) are called *stationary* in κ, and the dual ideal to the club filter is called the ideal of *non-stationary* sets (so a subset of κ is non-stationary if it has empty intersection with some club set).

THE AXIOM OF CHOICE

Show that club sets are stationary, and that bounded sets are non-stationary (hence stationary sets are unbounded in κ). If $\operatorname{cf}\kappa = \lambda$ and $\mu < \lambda$ is a regular cardinal, show that the set $\{\delta < \kappa \mid \operatorname{cf}\delta = \mu\}$ is stationary (in κ).

[An ascending sequence of length μ in a club set X must have a limit of cofinality μ.]

Club sets are also closed under *diagonal intersections*. For κ regular, given κ subsets $\langle C_\alpha \mid \alpha < \kappa \rangle$ of κ, the diagonal intersection $\Delta \langle C_\alpha \mid \alpha < \kappa \rangle$ is $\{\gamma < \kappa \mid \forall \delta < \gamma (\gamma \in C_\delta)\}$.

Show that $\Delta \langle C_\alpha \mid \alpha < \kappa \rangle = \bigcap_{\alpha < \kappa}(C_\alpha \cup \alpha + 1)$ and that if κ is regular and each C_α is club, then $\Delta \langle C_\alpha \mid \alpha < \kappa \rangle$ is also club.

More generally if $\operatorname{cf}\kappa = \lambda < \kappa$ we can define diagonal intersections along any $f : \lambda \to \kappa$ which is increasing and cofinal in κ: define

$$\Delta_f \langle C_\alpha \mid \alpha < \lambda \rangle = \{\gamma < \kappa \mid \forall \delta < \lambda (f(\delta) < \gamma \Rightarrow \gamma \in C_\delta)\}.$$

Again show that if each C_α is club, then $\Delta_f \langle C_\alpha \mid \alpha < \lambda \rangle$ is also club.

[Define an increasing sequence $\langle \gamma_n \mid n < \omega \rangle$ such that $\gamma_{n+1} \in \bigcap \{C_\delta \mid f(\delta) < \gamma_n\}$ and let $\gamma_\omega = \bigcup \gamma_n$, and show that if $f(\delta) < \gamma_\omega$ then $\gamma_\omega \in C_\delta$. Hence $\gamma_\omega \in \Delta_f \langle C_\alpha \mid \alpha < \lambda \rangle$.]

Now the reason for the name *stationary* comes from work of Fodor [Fod56], involving regressive functions: a function $f : X \to \kappa$, where $X \subseteq \kappa$, is *regressive* if $f(\alpha) < \alpha$ for all $\alpha \in X$, $\alpha \neq 0$. First show that if $\operatorname{cf}\kappa > \omega$ then no regressive function can *diverge* on a club $X \subseteq \kappa$. That is, if f is regressive with domain X club in κ, then the values $f(\alpha)$ cannot tend to κ in the standard sense: we cannot have $\forall \beta < \kappa \exists \gamma < \kappa \forall \delta (\gamma < \delta \in X \Rightarrow f(\delta) > \beta)$.

[The contrapositive, which is what we want to generalize, is that for any such f, there must be some $\beta < \kappa$ such that the set $\{\delta \in X \mid f(\delta) \leq \beta\}$ is unbounded. If not, we could define an increasing function $g : \omega \to X$ such that $g(n+1) < \delta \in X \Rightarrow f(\delta) > g(n)$ for each n. Then $g(\omega) = \bigcup_{n<\omega} g(n)$ would be in X (since X is closed) with $f(g(\omega)) \geq g(n)$ for each $n < \omega$, i.e. f cannot be regressive at $g(\omega)$.]

Now Fodor's theorem states:

Theorem 7.4.7 *(AC)* If S is stationary in a regular cardinal κ, $\kappa > \omega$ and $f : S \to \kappa$ is regressive, then f is constant on some stationary set.

[Prove this by assuming otherwise. Take a counterexample $f : S \to \kappa$. For each $\alpha < \kappa$ choose a club C_α with $f^{-1}(\alpha) \cap C_\alpha = \emptyset$. (The axiom of choice is needed here.)

Now let $D = \Delta \langle C_\alpha \mid \alpha < \kappa \rangle$, and take $\gamma \in S \cap D$ (which is not empty since S is stationary and D is club). Then $f(\gamma) < \gamma$ so since $\gamma \in D$, we have $\gamma \in C_{f(\gamma)}$; but this contradicts the choice of $C_{f(\gamma)}$.]

Show that Fodor's theorem cannot be proved in exactly the same form for singular κ with uncountable cofinality; but we can show that a function regressive on a stationary set must then be *bounded* on a stationary set.

[If $\lambda = \mathrm{cf}\,\kappa < \kappa$ then κ must have a club set C_0 of order type λ (say $C_0 = \{\gamma_\delta \mid \delta < \lambda\}$); we can intersect the given stationary set S with C_0, and get a regressive function on a stationary subset of λ by considering those $\delta < \lambda$ for which $f(\gamma_\delta) < \gamma_{\delta'}$ for some $\delta' < \delta$, and taking $g(\delta) = \delta'$. This must happen at least for all limit δ, which is a club subset of λ. So Fodor's theorem for λ will give the result as stated.

To see that no more is possible, consider $\kappa = \aleph_{\omega_1}$ and $S = C_0 = \{\aleph_\alpha \mid \alpha < \omega_1\}$, and take $f(\aleph_\alpha) = \alpha$.]

(9) The axiom of determinacy (sometimes called *determinateness*). This involves conceptual *games* for two players, whom we shall call players I and II. They pick elements alternately from a set X, forming a sequence $(x_n) \in {}^\omega X$; player I picking the even elements, II picking the odd ones, in turn. If A is a given subset of ${}^\omega X$, the game $G(A)$ will be *won* by I if $(x_n) \in A$ and by II if $(x_n) \notin A$. Who will win?

A *strategy* for either player is simply a function $\sigma : {}^{<\omega}X \to X$, and we say that a player plays according to strategy σ if that player plays $\sigma(a)$ at each turn, where $a \in {}^{<\omega}X$ is the sequence of previous choices. Then we say that σ is a *winning strategy* for player I if I wins when playing according to σ, for every possible play by II; and similarly for a winning strategy for player II. The basic question then is whether there exists a winning strategy for either player; if there is, then the game $G(A)$ is said to be *determined*.

The *axiom of determinacy* for X is then the statement that every game $G(A)$ for $A \subseteq {}^\omega X$ is determined. Show that for $X = 2$ or $X = \omega$, this contradicts the axiom of choice; i.e. use the axiom of choice to give sets $A \subseteq {}^\omega 2$ and $A' \subseteq {}^\omega \omega$ for which there is no winning strategy for either player.

[The simplest proof just uses the fact that in both cases, the set of all strategies and the set of all possible plays are both sets of the size of the continuum, and we use the axiom of choice to well-order the continuum and construct the sets A and A' by induction over this well-ordering.

Given a strategy σ and a possible play (an ω-sequence) τ, let $\sigma^{\mathrm{I}} * \tau$ be the result of player I playing according to σ against II playing τ; and let $\sigma^{\mathrm{II}} * \tau$ be the result of player II playing according to σ against I playing τ. Let $(\sigma_\alpha)_{\alpha<\mathfrak{c}}$ be an enumeration of all strategies, and $(\tau_\alpha)_{\alpha<\mathfrak{c}}$ an enumeration of all possible plays, where $\mathfrak{c}$ is the cardinal of the continuum. Then to construct set A, at stage α find the first τ_β and $\tau_{\beta'}$ such that $\sigma^{\mathrm{I}} * \tau_\beta$ and $\sigma^{\mathrm{II}} * \tau_{\beta'}$ have not yet been considered, and put $\sigma^{\mathrm{II}} * \tau_{\beta'}$ into A and $\sigma^{\mathrm{I}} * \tau_\beta$ into the complement of A.]

Show further that the assumption of a non-principal ultrafilter on ω

THE AXIOM OF CHOICE

implies that the axiom of determinacy for ω is false.

[Translate a sequence $(x_n) \in {}^\omega\omega$ into a subset Y of ω by taking the first x_0 numbers as being in Y, the next x_1 as being out of Y, the next x_2 in, and so on; then let (x_n) be a win for I if and only if $Y \in \mathcal{U}$ where $\mathcal{U}$ is the given non-principal ultrafilter on ω. Show that no σ can be a winning strategy for either player. Consider the following two plays (a) and (b) by II against strategy σ by I: in both, I plays first, and plays $s_0 = \sigma(\emptyset)$. In (a), II plays 0 and I replies with $s_1 = \sigma(s_0, 0)$; while in (b) II plays s_1 as first move and awaits the reply $s_2 = \sigma(s_0, s_1)$ from I. Then II plays s_2 as play three in (a), and awaits reply s_3 which is then played as play three in (b), and so on alternately. The resulting two plays, when translated into subsets of ω as suggested, will be complements except for the first s_0 members, and so one will be in $\mathcal{U}$ and the other will not.

Despite this contradiction of the axiom of choice, the axiom of determinacy is of considerable interest. One basic direction of interest is to restrict attention to certain collections of subsets of ${}^\omega X$ for which the assumption of determinacy does not contradict the axiom of choice; most of the collections studied in *descriptive set theory* (see [Kec95], [Mos80]) will be of this sort, and determinacy turns out to be an important property for those collections which have it. Another direction of interest is to consider sub-universes of sets within which the axiom of choice is false but the axiom of determinacy true. For some years the consistency strength of the theory ZF + axiom of determinacy was open, and its classification in terms of large cardinals turned out to be an important advance in the study of large cardinals (see [Kan94]).]

8
Constructible sets and forcing

The continuum hypothesis of Cantor was the first question asked in set theory, which (as it turned out) could not be answered from the axioms which have come to be accepted (i.e. from ZFC, assuming ZFC to be consistent). Paul Cohen introduced the notion of *forcing* to complete the proof of this in 1963, [Coh63a], Gödel having earlier introduced the notion of *constructible sets* to show that it could not be disproved. In essence, Gödel gave a model which was very thin, in the sense that it contained very few subsets of each infinite set (in fact as few as possible, given that it should still give a model of ZF). He was able to show that in this model, the generalized continuum hypothesis and the axiom of choice both hold. Then Cohen gave a method for adding more subsets to some infinite set (originally to ω), and then completing the construction so that he still had a model of ZF or of ZFC. His original presentation was in terms of adding new subsets to Gödel's model, but it was then generalized to a method of adding new subsets to any suitable model.

Cohen showed that the continuum hypothesis was false in one of his models, and Solovay and Easton, [Sol65] and [Eas70], then showed that the results sketched in the exercises in 6.7 imply all that can be proved from ZFC about cardinal exponentiation, in the case of regular cardinals. In effect, for regular cardinals κ the value of 2^κ is almost unconstrained; it has to be $\geq 2^\mu$ for any $\mu < \kappa$, and it has to have cofinality $> \text{cf}(\kappa)$, and that is all we can prove about it. Even the addition of axioms of extent as mentioned in 9.4 has little effect.

Cohen also gave a model of ZF in which the axiom of choice fails. Since Cohen's work, very many other questions have been shown to be independent of ZF or of ZFC; i.e. neither the statement nor its negation can be proved from those axioms. We present in this chapter some of this work, starting with Gödel's.

8.1 Gödel's constructible sets

We give some details here of the method introduced by Gödel in [Göd38] to give his proof that the continuum hypothesis could *not* be disproved from ZFC, assuming ZF is consistent. This method, introducing the *Gödel constructible sets*, was an important ingredient in the subsequent development of independence proofs by Cohen and others.

What Gödel did was to define a collection of sets, L, which he called constructible. L is in fact a proper class and contains all ordinals, and it is sufficiently closed under ordinary operations of set theory (such as unions, pair sets, images of functions), that it can in fact be proved to be a model of all of the axioms of ZFC and further even of the generalized continuum hypothesis (the GCH) as introduced in 6.7. But the definition and the properties of L do not need the axiom of choice or the GCH; they need only the axioms of ZF. As a result, we have a relative consistency result: if ZF is consistent, then so is ZFC plus GCH. In fact we could go further, with Gödel, and introduce the *axiom of constructibility* which is simply the statement that every set is constructible in Gödel's sense; this is usually referred to as $V = L$ (which is how it can be written in a theory which allows class variables, such as VNB in 9.2, which was used for just this purpose). The main results then take the form $V = L \Rightarrow$ AC and $V = L \Rightarrow$ GCH.

8.1.1 The definition of L

We give here the most straightforward definition. The definition is by transfinite induction, and superficially looks similar to the definition of V as in 6.6.13(3). But the full power-set operation, which was iterated to produce V, is here replaced by a "definable power-set" which will in general be very much smaller. If X is any transitive set, we form the language $\mathcal{L}_X$ which is the language of set theory with constant symbols $\overset{\circ}{x}$ added to denote each member x of X. Then for each formula $\varphi(v)$ of $\mathcal{L}_X$ with just one free variable v, say that $\varphi(v)$ defines the subset $\{x \in X \mid (X, \in) \models \varphi(\overset{\circ}{x})\}$. (Note that $\varphi(\overset{\circ}{x})$ will have no free variables, so no assignment is needed in the satisfaction relation here.) The collection of all such subsets, defined by all the formulas of $\mathcal{L}_X$ with just one free variable, is the definable power-set of X, which we shall denote by $\mathcal{D}(X)$. So we are allowing parameters from X into our definitions, and this will mean for example that all finite subsets of X will count as definable; we are using a common meaning for definable, but it is not the only one, and we may emphasize what we are doing here by saying *definable with parameters from X*.

This can be described quite quickly; but we shall see that the important point is that this can all be done *within* set theory and with relatively simple

CONSTRUCTIBLE SETS AND FORCING   129

definitions. We show this first.

First we give a representation of the language $\mathcal{L}_X$ within set theory. This can be thought of as we thought of the representation of numbers in set theory in chapter 6; indeed, Gödel gave a representation of a first-order language within arithmetic when presenting his incompleteness theorems, and we could consider the representation of numbers in set theory as carrying over Gödel's representation of the first-order language. But we shall in fact start from scratch, so that we can present more of the details, and make use of the structure of sets directly.

What we shall do is to define, for each formula φ of $\mathcal{L}_X$, a corresponding set $\ulcorner \varphi \urcorner$, which we shall refer to as the *Gödel-set* of φ (by analogy with Gödel-numbers). (The symbols $\ulcorner$ and $\urcorner$, usually called *corners*, are borrowed from Quine.) We first need terms $\ulcorner \overset{\circ}{x} \urcorner$ for constant terms $\overset{\circ}{x}$ and $\ulcorner v_i \urcorner$ for variables; and for each symbol s we shall write $\ulcorner s \urcorner$ for a set to correspond to that symbol; then $\ulcorner \varphi \urcorner$ will simply be the finite sequence of the sets corresponding to the main symbol and main subformulas of φ (or just to the symbols of φ when φ is atomic).

Definition 8.1.2 First, the symbols:

(i) $\langle x, 0 \rangle$ for $\ulcorner \overset{\circ}{x} \urcorner$ (for $x \in X$);
(ii) $\langle i, 1 \rangle$ for $\ulcorner v_i \urcorner$ (for $i \in \omega$);
(iii) $\langle 0, 2 \rangle$ for $\ulcorner = \urcorner$;
(iv) $\langle 1, 2 \rangle$ for $\ulcorner \in \urcorner$;
(v) $\langle 2, 2 \rangle$ for $\ulcorner \neg \urcorner$;
(vi) $\langle 3, 2 \rangle$ for $\ulcorner \Rightarrow \urcorner$;
(vii) $\langle 4, 2 \rangle$ for $\ulcorner \exists \urcorner$;
(viii) $\langle 5, 2 \rangle$ for $\ulcorner [\urcorner$; and
(ix) $\langle 6, 2 \rangle$ for $\ulcorner] \urcorner$.

(The choice of these representatives is not going to be important, we just want simple sets which we know are distinct.)

Next, for terms and formulas: it turns out to be easier to work with the whole build-up of the formula from symbols, rather than just the sequence of symbols. So we work with the complete sequences of subformulas of the language $\mathcal{L}_X$, which is described by FmlSeq below:

(x) $\mathrm{Term}(x, X)$ for $\exists i \in \omega (x = \ulcorner v_i \urcorner) \vee \exists y \in X (x = \ulcorner \overset{\circ}{y} \urcorner)$
    (x is a term in $\mathcal{L}_X$);
(xi) $\mathrm{Atfml}(x, X)$ for $\mathrm{Func}(x) \wedge \mathrm{dom}(x) = 3 \wedge \mathrm{Term}(x(0), X) \wedge$
    $\wedge [x(1) = \ulcorner = \urcorner \vee x(1) = \ulcorner \in \urcorner] \wedge \mathrm{Term}(x(2), X)$
    (x is an atomic formula of $\mathcal{L}_X$, or rather the Gödel-set of an atomic formula);
(xii) $\mathrm{Nfml}(x, y)$ for $\mathrm{Func}(x) \wedge \mathrm{dom}(x) = 2 \wedge x(0) = \ulcorner \neg \urcorner \wedge x(1) = y$

(x is the negation of the formula y, or rather x is the Gödel-set of the negation of the formula whose Gödel-set is y; but we shall not always make these nice distinctions in future);

(xiii) Ifml(x, y, z) for Func$(x) \wedge \operatorname{dom}(x) = 5 \wedge$
$$\wedge x(0) = \ulcorner [\urcorner \wedge x(1) = y \wedge x(2) = \ulcorner \Rightarrow \urcorner \wedge x(3) = z \wedge x(4) = \ulcorner] \urcorner$$
(x is the implication: if y then z);

(xiv) Efml(x, y) for Func$(x) \wedge \operatorname{dom}(x) = 3 \wedge$
$$\wedge x(0) = \ulcorner \exists \urcorner \wedge \exists i \in \omega (x(1) = \ulcorner v_i \urcorner) \wedge x(2) = y$$
(x is an existential quantification of formula y);

(xv) FmlSeq(u, x, n, X) for

$$\operatorname{Func}(x) \wedge \operatorname{dom}(x) = n + 1 \wedge x(n) = u \wedge n \in \omega \wedge$$
$$\wedge \forall k < n + 1 [\operatorname{Atfml}(x(k), X) \vee \exists j, l < k$$
$$[\operatorname{Nfml}(x(k), x(j)) \vee \operatorname{Ifml}(x(k), x(j), x(l)) \vee \operatorname{Efml}(x(k), x(j))]]$$

(x is a complete description of the build-up of u from all of its subformulas, and ultimately from atomic formulas, in the language $\mathcal{L}_X$).

We can now define formulas, if we need to, by Fml(u, X) for

$$\exists x \exists n < \omega \operatorname{FmlSeq}(u, x, n, X)$$

but we shall see that it is usually easier to work with the sequence describing the build-up. Note that we have, by this definition, defined a formula not as just a sequence of symbols, but rather as a structure which is closer to what is usually called the *parse tree* of the formula. We have said that a formula is always a sequence or function of length three or five; and for a complex formula, the individual variables or constant terms which occur may be buried quite deep in its structure. But the formula sequence describing the build-up of that formula *will* have all the variables and constant terms showing at the top level, since they must occur in an atomic formula at some stage in the build-up. (These details could be varied in many ways, and it is a good exercise to give alternative definitions which may be closer to the interpretation of formulas as sequences of symbols, and then consider the resulting changes needed in the subsequent definitions.)

Now having our representation of the language, we can represent the satisfaction of formulas in a structure, and hence define the definable power-set operation.

The structures we shall be using will be standard transitive structures which are just transitive sets, with the standard membership relation. We shall take them to be the sets X we have been using for our language. What we shall define are then the sets of *satisfying assignments* for formulas in these structures. These assignments (which formally assign members of X to variables) can be taken to be functions from some natural number into X, since the variables are in one-to-one correspondence with the natural numbers, and any formula has only a finite number of free variables, and hence uses only a finite number of values of any assignment.

To give this definition we parallel the definition of the sequence of subformulas which builds the given formula, defining for each of the subformulas its set of satisfying assignments. We first need a measure of which variables can occur in the formula. To start with we won't need to know whether they occur free or bound, though we clearly could give such definitions if needed; the definition we give will take care of that automatically. We use this measure to limit the length of the assignments. (There will be good reasons for thinking only of finite sequences, not infinite ones, so we shall want limited assignments of this sort.) Then for each subformula in turn (starting with the atomic ones) we define the set of all assignments of the agreed length which satisfy that subformula. The definitions are all given assuming that the set x is a formula sequence, as above; it will not matter what they give in other circumstances.

Definition 8.1.3 (i) $\text{Vlength}(x)$ for $1 + \sup\{n \in \omega \mid \exists i(\text{Atfml}(x(i), X) \wedge \wedge (x(i)(0) = \ulcorner v_n \urcorner \vee x(i)(2) = \ulcorner v_n \urcorner)\}$

(one more than the highest variable's suffix that appears in x; note how it might be nicer to use the suffix notation x_i for the ith value of function x, and then $x_i(0)$ for the first value of that value, instead of $x(i)(0)$, etc.).

Now each subformula appearing in x will in turn be given its set of satisfying assignments by a function S of the same length as x. The assignments themselves will be functions with domain $\text{Vlength}(x) = r$, say, and so members of ^rX. If v_n is the highest variable, then $n \in r = n+1$, which is why the $+1$ appears.

The effect of the definition will be that if $x(i) = \ulcorner \varphi \urcorner$, then $S(i)$ must be the set $\{a \in {}^rX \mid (X, \in) \models \varphi(a)\}$, where $r = \text{Vlength}(x)$, so such an a will assign $a(n)$ to v_n.

So for $x(i)$ atomic, suppose $x(i) = \ulcorner v_j \in v_l \urcorner$. Then we must have $S(i) = \{a \in {}^rX \mid a(j) \in a(l)\}$. If $x(i) = \ulcorner v_j \in \overset{\circ}{z} \urcorner$ (where $z \in X$) then we must have $S(i) = \{a \in {}^rX \mid a(j) \in z\}$; if $x(i) = \ulcorner \overset{\circ}{z} \in v_j \urcorner$ (where $z \in X$) then we must have $S(i) = \{a \in {}^rX \mid z \in a(j)\}$. And if $x(i) = \ulcorner \overset{\circ}{y} \in \overset{\circ}{z} \urcorner$ (with $y, z \in X$) we must have $S(i) = {}^rX$ if $y \in z$ and $S(i) = \emptyset$ if $y \notin z$. There will be four similar clauses for atomic formulas with $=$ instead of $\in$; the first would be: if $x(i) = \ulcorner v_j = v_l \urcorner$, we must have $S(i) = \{a \in {}^rX \mid a(j) = a(l)\}$, and the others we leave as exercises.

For compound formulas, if $x(i) = \ulcorner \neg x(j) \urcorner$ then we must have $S(i) = {}^rX - S(j)$. If $x(i) = \ulcorner x(j) \Rightarrow x(l) \urcorner$ then we must have $S(i) = ({}^rX - S(j)) \cup S(l)$ (remembering that $\varphi \Rightarrow \psi$ is equivalent to $\neg \varphi \vee \psi$). And if $x(i) = \ulcorner \exists v_j x(l) \urcorner$ then $S(i)$ must be the set of all j-variants of members of $S(l)$, which will be $\{a \in {}^rX \mid \exists b \in S(l)[\forall m < r(m \neq j \Rightarrow b(m) = a(m))]\}$.

(This gives eleven clauses which effectively repeat the definition of satisfaction from 2.4.1, and they should be compared with that definition to be sure that the reader understands both that definition and this. Note that no constant terms $\overset{\circ}{x}$ were included in the earlier definition.)

So formally we can define the condition:

(ii) $\operatorname{SatSeq}(S, x, X)$ for

$$\exists u, n, r(\operatorname{FmlSeq}(u, x, n, X) \land$$
$$\land \operatorname{Func}(S) \land \operatorname{dom}(S) = n + 1 \land \operatorname{Vlength}(x) = r \land \forall k < n + 1$$
$$[\forall j, l \in \omega(x(i) = \ulcorner v_j \in v_l \urcorner \Rightarrow S(i) = \{a \in {}^r X \mid a(j) \in a(l)\}) \land$$
$$[\forall j \in \omega \forall z \in X(x(i) = \ulcorner v_j \in \overset{\circ}{z} \urcorner \Rightarrow S(i) = \{a \in {}^r X \mid a(j) \in z\}) \land$$
$$\land \ldots]])$$

with nine further clauses in the final conjunction corresponding to the nine further clauses indicated above.

Now we can give a formal definition of satisfaction, i.e. the representation in ZF of $(X, \in) \models \varphi(a)$ (where a is an assignment to the variables up to the agreed length), by:

(iii) $\operatorname{Sat}(X, \ulcorner \varphi \urcorner, a)$ for $\exists x, S, n(\operatorname{SatSeq}(S, x, X) \land x(n) = \ulcorner \varphi \urcorner \land a \in S(n))$
(or alternatively for
$\forall x, S, n(\operatorname{SatSeq}(S, x, X) \land x(n) = \ulcorner \varphi \urcorner \Rightarrow a \in S(n))$;
we shall see some point in noting these sort of alternatives later).

The brief description of $\mathcal{D}(X)$ which we gave at the start of this section used formulas with just one free variable; and (although we could have done), we have not given the definition of free and bound variables in this representation of the language. The point is that we can get exactly the same sets by taking *projections* of the sets of satisfying assignments. If $A \subseteq {}^r X$, where $r \in \omega$, then for $i < r$ we define the i-projection of A, $\operatorname{proj}_i(A)$, by

$$\operatorname{proj}_i(A) = \{a(i) \mid a \in A\};$$

and for $i \geq r$ let $\operatorname{proj}_i(A) = \emptyset$. Then if A is any set of satisfying assignments, say $A = \{a \in {}^r X \mid \operatorname{Sat}(X, \ulcorner \varphi \urcorner, a)\}$, then for each $i < r$, $\operatorname{proj}_i(A)$ will be $\{x \in X \mid (X, \in) \models \varphi_i(\overset{\circ}{x})\}$ where $\varphi_i(\overset{\circ}{x})$ is the formula obtained from φ by prefacing it with $\exists v_j$ for each variable v_j which occurs free in φ with $j \neq i$, and substituting $\overset{\circ}{x}$ for free occurrences of v_i. (To see this is another nice exercise in checking the definition of satisfaction, 2.4.1, and particularly the clause for the existential quantifier.)

So finally we can define $\mathcal{D}(X)$ as:

CONSTRUCTIBLE SETS AND FORCING

Definition 8.1.4 $\mathcal{D}(X)$ for $\{\text{proj}_i(S(j)) \mid i, j \in \omega \land \exists x\, \text{SatSeq}(S, x, X)\}$.

Note that we have been liberal here, and have not attempted to limit the projections we take to those which are really intended. But if $i \geq r$ where $A \subseteq {}^r X$ then the definition of $\text{proj}_i(A)$ will give $\emptyset$; similarly if $j > n$ where S is a satisfaction sequence of length n, then $S(n) = \emptyset$; and our definition will certainly include $\emptyset$ in $\mathcal{D}(X)$ many times over. (Similarly it will include X many times over.) But in fact any way of defining $\mathcal{D}(X)$ which uses all formulas will include *every* definable subset infinitely many times over, since for any formula there are always infinitely many other formulas which are equivalent. So we may as well be liberal, provided we do not let in any subsets which are not definable.

We shall need some simple properties of $\mathcal{D}(X)$:

Lemma 8.1.5 *If X is finite, so is $\mathcal{D}(X)$, and in fact $\mathcal{D}(X) = \mathcal{P}(X)$. If X is infinite, then $|X| = |\mathcal{D}(X)|$. And if X is transitive, then $X \subseteq \mathcal{D}(X)$.*

Proof For X finite, this follows when we see that if $y \subseteq X$, say $y = \{x_1, \ldots, x_k\}$, then the formula

$$v_0 = \mathring{x}_1 \lor \ldots \lor v_0 = \mathring{x}_k$$

will define y as a member of $\mathcal{D}(X)$. For X infinite, we use the fact that the number of formulas in $\mathcal{L}_X$ will also be $|X|$; and for each $x \in X$, we can define $\{x\} \in \mathcal{D}(X)$ by the formula $v_0 = \mathring{x}$. And if X is transitive, then for $x \in X$ the formula $v_0 \in \mathring{x}$ will define x. □

8.2 The definition of L

These preliminaries allow us to give the definition of L, the class of all constructible sets:

Definition 8.2.1

$$L_0 = \emptyset; \quad L_{\alpha+1} = \mathcal{D}(L_\alpha); \quad L_\lambda = \bigcup_{\alpha < \lambda} L_\alpha \text{ for limit } \lambda;$$

$$\text{and} \quad L = \bigcup_{\alpha \in \text{On}} L_\alpha.$$

If we want to keep the definition within ZF, and avoid talk of classes, we take L as a predicate (or formula), and define

$$L(x) \Leftrightarrow \exists \alpha (x \in L_\alpha).$$

Many of the properties of these stages L_α are quite straightforward. For example:

Lemma 8.2.2 *(i) For α finite or $\alpha = \omega$, $L_\alpha = V_\alpha$; for α infinite, $|L_\alpha| = |\alpha|$.*
(ii) Each L_α is transitive.

Proof Both follow by 8.1.5, by transfinite induction. □

But the deeper properties involve the fact that these stages are *absolute*; the definitions can be given in a sufficiently simple way, so that it makes no difference if they are interpreted in V or in L or in suitable L_α's; we shall get the same sets in each case.

This notion of interpreting formulas in different structures uses the notion of *relativized* formulas and terms:

Definition 8.2.3 We define for terms t and formulas φ of a language $\mathcal{L}_X$ the relativizations to a set or class Y, t^Y and φ^Y, where $X \subseteq Y$, by:

(i) if t is a variable v_i or constant term $\overset{\circ}{x}$ then t^Y is just v_i or $\overset{\circ}{x}$ again;
(ii) if t is $\{v_i \mid \varphi\}$ then t^Y is $\{v_i \in Y \mid \varphi^Y\}$;
(iii) if φ is an atomic formula $t \in u$ or $t = u$ for terms t and u, then φ^Y is $t^Y \in u^Y$ or $t^Y = u^Y$ respectively;
(iv) if φ is $\neg\psi$ or $[\psi \Rightarrow \chi]$ then φ^Y is $\neg\psi^Y$ or $[\psi^Y \Rightarrow \chi^Y]$ respectively;
(v) if φ is $\exists v_i \psi$ then φ^Y is $\exists v_i \in Y(\psi^Y)$.

Note that the typical Y in this definition will be L_α or L, so we shall want both the case where Y is a set and the case where Y is a proper class. The definition is fine as it stands for Y a set; for Y a proper class, we should rewrite all occurrences of $v_i \in Y$ as $Y(v_i)$, in other words we treat proper classes as circumlocutions for formulas with one free variable, as we did for L in 8.2.1. With this circumlocution, relativization to V (where V is the universe of all sets) is redundant, which is what we would want: for any formula, $\varphi \Leftrightarrow \varphi^V$ and for any term, $t = t^V$.

The relativized forms of terms and formulas will in effect be those terms or formulas interpreted in Y, and we can now give the definition of absoluteness:

Definition 8.2.4 Suppose that φ is a formula of $\mathcal{L}_X$ with the variables x_1, $x_2, \ldots, x_k$ free, and that $X \subseteq Y$. Then φ is *absolute between X and Y* if

$$\forall x_1, x_2, \ldots, x_k (x_1 \in X \wedge x_2 \in X \wedge \ldots \wedge x_k \in X \Rightarrow (\varphi^X \Leftrightarrow \varphi^Y)).$$

A term t of $\mathcal{L}_X$ is absolute between X and Y if the formula $v_i = t$ is absolute between X and Y (where v_i is a variable not appearing in t).

This says that provided the free variables of the formula or term are in X, and the constant symbols that appear refer to members of X (which will be

CONSTRUCTIBLE SETS AND FORCING

so if it is in the language with parameters from X), the formula or term will mean the same whether it is interpreted in X or in Y.

If a formula or term is absolute between X and V, we say that it is *absolute for X*; and if it is absolute for any *transitive* X for which the formula or term is in $\mathcal{L}_X$, we shall just say that it is *absolute*.

To see the point of this, notice that it certainly is not true for the power-set operation. The set ω will be absolute (i.e. there is a term describing it which is absolute), so it will mean the same in L as in V, and in any L_α for $\alpha > \omega$. (We shall show this in more detail later.) But the power-set of ω will not be absolute. If we interpret $\mathcal{P}(\omega)$ in V (which is the usual way we think of it), then we get every subset of ω. If we interpret $\mathcal{P}(\omega)$ in L, it can be shown that we shall get every Gödel-constructible subset of ω. Note that this will not just be those subsets of ω which are in the definable power-set of L_ω, though many will arise at that level. It can be shown that more subsets of ω will become definable, at some further levels L_α, for arbitrarily large ordinals α as long as these α are still countable in L. And for many of these α it will make perfect sense to interpret $\mathcal{P}(\omega)$ (they can be models of as many of the axioms of ZFC as we wish, though we may not be able to show that any are models of all of ZFC); and the interpretation of $\mathcal{P}(\omega)$ within such an L_α, is in fact a countable set. This will be just the property we shall use in order to show that the continuum hypothesis will hold in L; the cardinal of the power-set of ω as interpreted in L will have the property that any smaller cardinal is countable, and so that cardinal must count as $\aleph_1$ in L. But the continuum hypothesis may well be false in V, in which case the interpretation of $\mathcal{P}(\omega)$ must have changed. And note that even if the interpretation did not change between V and L, it certainly must change between V and L_α for any countable α since within L_α it will be interpreted by a set which is in fact countable.

In general it will be the case that any notion which can give uncountable cardinals, such as power sets, or cardinals themselves, or cofinality, or function sets ^XY with X infinite, will not be absolute. But a careful analysis will be needed to show that a notion *is* absolute, and this is the main part of the work in establishing Gödel's results on constructible sets. Similar work is needed to establish the basis for forcing.

8.2.5 *Exercises*

(1) Show that if $X \subseteq Y \subseteq Z$, and φ is absolute between X and Y and between Y and Z then it is absolute between X and Z; and if φ is absolute between X and Z and between Y and Z then it is absolute between X and Y.

(2) First deal with formulas without abstraction terms: show that atomic formulas are absolute if they have no abstraction terms.

Next show that formulas which use only bounded quantifiers and with no abstraction terms are absolute.

[Here we must use the fact that we assume X is transitive when defining *absolute* above. A bounded quantifier $\exists u \in v\ldots$ or $\forall u \in v\ldots$ can only refer to elements of X, if we assume that v is assigned in X (in other words the u must also be assigned in X, since X is transitive). So relativizing to X will give an equivalent formula.]

Hence show that the only way that absoluteness can fail for formulas without abstraction terms is that we have a formula with a quantifier (necessarily unbounded) of the form $\forall v\ldots$, which holds for all assignments of v in X but fails for some outside X, or (dually) of the form $\exists v\ldots$, which fails for all assignments in X but holds for one or more outside.

[We shall expand on this later in 8.6 when we introduce Skolem functions.]

(3) Absoluteness for a theory. A formula φ without abstraction terms is called Σ_1 if it is of the form $\exists y_1, \ldots, y_k \psi$ where ψ is bounded; and Π_1 if it is of the form $\forall y_1, \ldots, y_k \psi$ where ψ is bounded.

If we are given a theory T (which may be ZF, but it may be any theory), a formula φ is called Σ_1^T if it is provably equivalent in T to a Σ_1 formula, i.e. if there is a formula ψ which is Σ_1 such that

$$T \vdash \forall x_1, \ldots, x_n (\varphi \Leftrightarrow \psi)$$

where $x_1, \ldots, x_n$ are all the free variables of either φ or ψ. Π_1^T is defined similarly, and a formula is Δ_1^T if it is both Σ_1^T and Π_1^T.

Show that any formula which is Δ_1^T is absolute for any transitive model of T.

[We shall not usually be in a position to make much use of this method of proving absoluteness, since we shall not know in advance that the sets we are dealing with are models of ZF, or of enough of ZF to prove the equivalences that are needed. But we have noted in various places that such equivalents do exist, since they can be useful.]

(4) Abstraction terms will count as absolute under definition 8.2.4 if the term does not exist, i.e. if there is no element of X which is $\{x \mid \varphi\}$ then it will count as absolute ($v_i = \{x \mid \varphi\}$ will simply be false for all $v_i \in X$). But this is not really of concern; we shall be concerned with relatively simple terms and large enough sets X so that the terms will exist.

Show that if $\varphi(x, x_1, \ldots, x_k)$ is absolute for X and

$$\{x \in y \mid \varphi(x, x_1, \ldots, x_k)\} \text{ exists in } X \text{ whenever } y, x_1, \ldots, x_k \in X,$$

then $\{x \in y \mid \varphi(x, x_1, \ldots, x_k)\}$ is absolute for X. (Note that y may or may not be among $x_1, \ldots, x_k$.)

In fact we need more than these simple methods for the results concerning the L hierarchy. The following will suffice for what is needed:

CONSTRUCTIBLE SETS AND FORCING

Lemma 8.2.6 *(Further criteria for absoluteness)*

(i) *If the formula $\varphi(x, x_1, \ldots, x_k)$ is absolute between X and Y and is such that whenever $x_1, \ldots, x_k \in X$ and $\exists x \varphi(x, x_1, \ldots, x_k)$ holds in Y, there is an $x \in X$ such that $\varphi(x, x_1, \ldots, x_k)$ holds in Y, then $\exists x \varphi(x, x_1, \ldots, x_k)$ is absolute between X and Y.*

(ii) *If the term t and the formula φ (with free variables $x_1, \ldots, x_k$ in common), are both absolute between X and Y, and whenever $x_1, \ldots, x_k$ are in X, $t^Y \in X$, and further $\forall x_1, \ldots, x_k \in Y(\varphi^Y \Rightarrow x_1, \ldots, x_k \in X)$; then also the term $\{t \mid \varphi\}$ will be absolute between X and Y.*

Proof (i) $\exists x \varphi(x, x_1, \ldots, x_k)^X$ will be $\exists x \in X \varphi^X(x, x_1, \ldots, x_k)$, and the criterion given will be just what is needed for this to be the same as $\exists x \in Y \varphi^Y(x, x_1, \ldots, x_k)$ since we are given, for $x, x_1, \ldots, x_k \in X$, that $\varphi^X(x, x_1, \ldots, x_k) \Leftrightarrow \varphi^Y(x, x_1, \ldots, x_k)$.

Now for (ii) we note that $\{t \mid \varphi\}$ is an abbreviation for

$$\{v \mid \exists x_1, \ldots, x_k (v = t \wedge \varphi)\}.$$

So assume that this term gives u when interpreted in X, i.e. $u \in X$ and $(\overset{\circ}{u} = \{t \mid \varphi\})^X$. This latter formula is equivalent to

$$(\forall v(v \in \overset{\circ}{u} \Leftrightarrow \exists x_1, \ldots, x_k(v = t \wedge \varphi)))^X,$$

and we must show

$$(\forall v(v \in \overset{\circ}{u} \Leftrightarrow \exists x_1, \ldots, x_k(v = t \wedge \varphi)))^Y.$$

This is the conjunction of

$$\forall v \in Y(v \in \overset{\circ}{u} \Rightarrow \exists x_1, \ldots, x_k(v = t \wedge \varphi))^Y, \quad \text{and}$$

$$\forall v \in Y(\exists x_1, \ldots, x_k(v = t \wedge \varphi))^Y \Rightarrow v \in \overset{\circ}{u}).$$

The first follows immediately from the absoluteness assumed for t and φ. The second uses the extra hypotheses: for any $v, x_1, \ldots, x_k \in Y$, if $(v = t \wedge \varphi)^Y$ then $x_1, \ldots, x_k \in X$, and hence $t^Y \in X$ and so $t^Y = t^X$ and will be in u. □

8.2.7 Exercises

(1) Show that the following terms and formulas are all absolute:

$$\{a, b\}, \langle a, b \rangle, 0, 1, 2, \ldots$$

$\text{Trans}(x), \text{Connex}(x), \text{Ord}(x), \text{Int}(x), x < y, \text{Suc}(x), \text{Lim}(x)$

$\text{Func}(x), \text{range}(x), \text{dom}(x), f(x), f''x, f \upharpoonright x, \bigcup x, x + 1, x \times y$.

[These can all be done using 8.2.5(2) and (4).]

(2) Show that the following terms are absolute:

$$\omega, \text{TC}(x), V_\omega, \bigcup_{\alpha < \beta} f(\alpha), {}^r\!A \text{ for } r < \omega.$$

[These need 8.2.6(i) and (ii).]

We use these methods in 8.4; first we need another important property of natural hierarchies.

8.3   Reflection principles

The replacement axiom is very powerful, and allows us to show that (in effect) every formula or term will be absolute for many stages of the cumulative hierarchy. We give a proof of this which applies to any hierarchy, so that we can apply it directly to the constructible hierarchy L as well as to V.

Theorem 8.3.1 *Let M be any hierarchy satisfying: $M = \bigcup_{\alpha \in \text{On}} M_\alpha$ where $M_\alpha \subseteq M_\beta$ for $\alpha < \beta$ and $M_\lambda = \bigcup_{\alpha < \lambda} M_\alpha$ for limit λ. (Here M may be a proper class, but we assume each M_α is a set.) Let φ be any formula of $\mathcal{L}_X$ with the variables $x_1, x_2, \ldots, x_k$ free, where $X \subseteq M_{\alpha_0}$ for some α_0.*

Then for arbitrarily large β, φ is absolute between M and M_β, i.e.

$$\forall \alpha \exists \beta > \alpha \forall x_1, \ldots, x_k \in M_\beta [\varphi^{M_\beta} \Leftrightarrow \varphi^M].$$

Proof   We first assume that φ is written out to eliminate all defined terms, and then arranged in equivalent form with all quantifiers at the beginning (in prenex form). So suppose φ is

$$Q_1 y_1 \ldots Q_j y_j \psi(x_1, \ldots, x_k, y_1, \ldots, y_j)$$

where each Q_i is either $\exists$ or $\forall$, and $\psi(x_1, \ldots, x_k, y_1, \ldots, y_j)$ has no quantifiers. We then prove the result by induction on j.

For $j = 0$ we are dealing with a formula with no quantifiers, and so a Boolean combination of atomic formulas. But relativization does not affect atomic formulas, when there are no abstraction terms present; this is why we assume they have been eliminated, and so this case is trivial.

Now assume that we have the result for formula φ and that we are going to add one further quantifier, and for simplicity we prove the result for $\exists x_1 \varphi$ ($\forall x_1 \varphi$ will then be treated similarly, and obvious changes will give the result for other variables).

First for given α, let g be a function on ordinals with $g(0) = \max(\alpha, \alpha_0)$, where $X \subseteq M_{\alpha_0}$, and for all β, $g(\beta + 1)$ is some ordinal δ above $g(\beta)$

CONSTRUCTIBLE SETS AND FORCING 139

for which M_δ reflects φ, i.e. $\forall x_1,\ldots,x_k \in M_\delta[\varphi^{M_\delta} \Leftrightarrow \varphi^M]$ holds; the induction hypothesis implies that g can be defined for all ordinals. We shall assume further that this function g is continuous, i.e. that for limit λ, $g(\lambda) = \bigcup_{\beta<\lambda} g(\beta)$; this is clearly true for the case of no quantifiers, and we shall include in the induction the proof that this can always be made to hold.

Next for each δ, given any elements $x_2,\ldots,x_k$ from M_δ, suppose that there is some element $x_1 \in M$ for which $\varphi^M(\mathring{x}_1,\mathring{x}_2,\ldots,\mathring{x}_k)$ holds. Then we let $f(x_2,\ldots,x_k)$ be the least ordinal γ such that this holds for some $x_1 \in M_\gamma$. If there is no such x_1, let $f(x_2,\ldots,x_k) = \delta$. Now we use the replacement axiom to see that $\{f(x_2,\ldots,x_k) \mid x_2,\ldots,x_k \in M_\delta\}$ is a set, and hence a subset of some δ'; let $h(\delta)$ be the least such $\delta' \geq \delta$.

Next we compose this function with g to get a function g' with the property that, for any γ, $g'(\gamma)$ will be a limit value of g (i.e. $g'(\gamma) = g(\lambda)$ for some limit ordinal λ), and its values are closed under h (i.e. if $\delta < g'(\gamma)$ then $h(\delta) < g'(\gamma)$). We do this by taking $h_1(\delta) = g(\gamma')$ for the first γ' such that $g(\gamma') > h(\delta)$, then iterating h_1 by $h_1^{n+1}(\delta) = h_1(h_1^n(\delta))$ for $n < \omega$ and letting $h_2(\delta) = \bigcup_{n<\omega} h_1^n(\delta)$. Now let g' enumerate the values of h_2. This will be a continuous function with the desired property, since if $\delta < g'(\gamma)$, say $g'(\gamma) = h_2(\gamma')$, then $\delta < h_1^n(\gamma')$ for some $n < \omega$ and so

$$h(\delta) < h(h_1^n(\gamma')) < h_1(h_1^n(\gamma')) = h_1^{n+1}(\gamma') < h_2(\gamma') = g'(\gamma)$$

as required.

We now show that if $\beta = g'(\gamma)$ for any γ then $\exists x_1 \varphi$ is absolute between M and M_β, i.e. $\forall x_2,\ldots,x_k \in M_\beta[(\exists x_1 \varphi)^{M_\beta} \Leftrightarrow (\exists x_1 \varphi)^M]$. For, given $x_2,\ldots,x_k \in M_\beta$, since β is a limit, we must have $x_2,\ldots,x_k \in M_\delta$ for some $\delta < \beta$ by the assumption on the hierarchy M. So if there is some $x_1 \in M$ for which $\varphi^M(\mathring{x}_1,\mathring{x}_2,\ldots,\mathring{x}_k)$ holds, then $x_1 \in M_{h(\delta)}$ and hence $x_1 \in M_\beta$. So $\exists x_1 \in M_\beta \varphi^{M_\beta}(x_1,\mathring{x}_2,\ldots,\mathring{x}_k)$ will hold (using φ is absolute between M and M_β), and so $\forall x_2,\ldots,x_k \in M_\beta[(\exists x_1 \varphi)^M \Rightarrow (\exists x_1 \varphi)^{M_\beta}]$ will hold. The other way round, $\forall x_2,\ldots,x_k \in M_\beta[(\exists x_1 \varphi)^{M_\beta} \Rightarrow (\exists x_1 \varphi)^M]$, will always hold when $M_\beta \subseteq M$ and φ is absolute between M and M_β, so we have the result.

The changes needed in this induction step to deal with the case $\forall x_1 \varphi$ are clear when we note that it will be the other way round which always holds for $\forall x_1 \varphi$, i.e. $\forall x_2,\ldots,x_k \in M_\beta[(\forall x_1 \varphi)^M \Rightarrow (\forall x_1 \varphi)^{M_\beta}]$ will always hold when $M_\beta \subseteq M$ and φ is absolute between M and M_β. So in defining the function f, this time we look for places where there is some element $x_1 \in M$ for which $\varphi^M(\mathring{x}_1,\mathring{x}_2,\ldots,\mathring{x}_k)$ fails, and ensure that if there is such an element, then there will be such an element in the set M_β. Then we shall have the dual case, $\forall x_2,\ldots,x_k \in M_\beta[(\forall x_1 \varphi)^{M_\beta} \Rightarrow (\forall x_1 \varphi)^M]$, and hence $\forall x_1 \varphi$ will be absolute between M and M_β. □

8.3.2 Exercises

(1) Many variants of the reflection principle proved above can be given; some will assume further properties of the hierarchy M. One useful variant is to show that we can reflect on any finite set of formulas together. Show that provided all hereditarily finite sets are in some M_α, then for any formulas $\varphi_1, \ldots, \varphi_m$, with at most the variables $x_1, x_2, \ldots, x_k$ free, for arbitrarily large β all of $\varphi_1, \ldots, \varphi_m$ are absolute between M and M_β.

[Reflect on the formula Φ:
$$((u = 1) \wedge \varphi_1) \vee ((u = 2) \wedge \varphi_2) \vee \ldots \vee ((u = m) \wedge \varphi_m)$$
where u is a new variable, using the fact that the formulas $u = 1$ etc. will all be absolute between any M and M_β if both contain all hereditarily finite sets. Why is it not sufficient simply to reflect on the formula $\varphi_1 \wedge \varphi_2 \wedge \ldots \wedge \varphi_m$?]

Hence show that there are arbitrarily good approximations to models of ZFC among the stages V_α of the cumulative hierarchy.

[Apply the above to the V_α hierarchy, with $\varphi_1, \ldots, \varphi_m$ any finite set of axioms of ZFC.]

(2) Show that we cannot expect to prove a reflection principle which reflects on infinitely many formulas at once.

[Apply such a principle to the V_α hierarchy to get stages of the cumulative hierarchy which are models of the whole of ZFC, and hence derive the consistency of ZFC. Of course, such an argument could be turned round into an argument for the plausibility of the existence of stages V_α which are models of ZFC, given the plausibility of this stronger reflection principle. We shall look at much stronger principles in 9.4. See [Lev60].]

8.4 Properties of L

We now use the reflection principle to show that L is a model of ZF; what we actually prove is that for each axiom φ of ZF, φ^L is a theorem of ZF. The first of these are simple:

Proposition 8.4.1 *If φ is one of the axioms of extensionality, foundation, null-set, pair-set, sum-set, or infinity, then φ^L is a theorem of ZF.*

Proof We simply note that in all these cases, we can write a formula φ with only bounded quantifiers, which implies the axiom; directly, in the case of extensionality and foundation, and because the appropriate sets will be in L in the other cases. Then φ will be absolute by 8.2.5(1), and the result will follow, since L is transitive.

CONSTRUCTIBLE SETS AND FORCING

For extensionality, the formula φ can be:

$$(\forall x \in a(x \in b) \land \forall x \in b(x \in a)) \Rightarrow a = b,$$

and this is equivalent to the axiom.

For foundation, φ can be $\exists x \in a(x = x) \Rightarrow \exists x \in a(\forall y \in x(y \notin a))$.

For the null-set axiom, φ can be $\forall y \in x(y \neq y)$. Since $\emptyset$ is in L it will satisfy this φ in L (since it does in V) and so the relativized axiom is proved.

For the pair-set axiom, φ can be

$$\forall y \in x(y = a \lor y = b) \land a \in x \land b \in x.$$

If a and b are both in L_α, then $\{a, b\}$ is defined in L_α by the formula $y = \overset{\circ}{a} \lor y = \overset{\circ}{b}$, and so is in $L_{\alpha+1}$ and so in L, and will satisfy φ in V and hence in L, so the relativized axiom is proved.

Similarly for the sum-set axiom, φ can be

$$\forall y \in x \exists z \in a(y \in z) \land \forall z \in a \forall y \in z(y \in x),$$

and $\bigcup a$ is defined by $\exists z \in \overset{\circ}{a}(y \in z)$.

For the axiom of infinity, we can write $\mathrm{Ord}\, y$ (i.e. $\mathrm{Trans}\, y \land \mathrm{Connex}\, y$) as the formula

$$\forall z \in y \forall u \in z(u \in y) \land \forall z \in y \forall u \in y(u \in z \lor u = z \lor z \in u)$$

and note that this formula, since it will be absolute, will define the set of all ordinals in any L_α. So by induction we can see that the set of all ordinals in L_α will be α itself, for each α, and $\alpha \in L_{\alpha+1}$ (it will be defined by this formula in L_α), and in particular $\omega \in L_{\omega+1}$ and hence $\omega \in L$. Then the axiom of infinity says $\exists w(\mathrm{Ind}(w))$ and since $\mathrm{Ind}(\omega)$ can be written with bounded quantifiers, we shall have the result. □

Note that the proof above shows that all ordinals are constructible (since $\alpha \in L_{\alpha+1}$) and hence L is a proper class and not a set (since it contains the proper class On).

For the remaining axioms, we take two steps: first we show that we can prove a weak form of the power-set and replacement axioms, which will give the full form when we prove also the subset axioms. Then we complete the result by using the reflection principle to prove the subset axioms.

First we need a definition:

Definition 8.4.2 For $x \in L$, $\mathrm{od}(x)$ is the least α such that $x \in L_\alpha$. This is the *constructible order* of x for any constructible set; note that since $L_0 = \emptyset$ and $L_\lambda = \bigcup_{\alpha<\lambda} L_\alpha$ for limit λ, $\mathrm{od}(x)$ will always be a successor ordinal $\beta + 1$. If needed we can let $\mathrm{od}(x)$ be 0 for $x \notin L$.

Proposition 8.4.3 *Let the weak power-set axiom be* $\exists x \forall y (y \subseteq a \Rightarrow y \in x)$, *and the weak replacement axiom be*

$$\forall z, u, v(\psi(z,u) \wedge \psi(z,v) \Rightarrow u = v) \Rightarrow \exists x \forall y (\exists z \in a \, \psi(z,y) \Rightarrow y \in x).$$

Then these hold in L; i.e. their relativizations to L are theorems of ZF.

Proof In both cases we use the fact that any set which is a subset of L must be a subset of L_α for some α. To see this, suppose $Y \subseteq L$ is a set, and let $\alpha = \bigcup_{z \in Y} \text{od}(z)$. Then α will be a set by the replacement and union axioms, (it is $\bigcup \{\beta \mid \exists z \in Y(\beta = \text{od}(z)\}$ and $\text{od}(z)$ is a function), and clearly $Y \subseteq L_\alpha$.

Now for the weak power-set axiom, since $y \subseteq a$ is a bounded formula and so absolute, $\{y \mid (y \subseteq a)^L\} \subseteq \mathcal{P}(a)$ so we take $\{y \mid (y \subseteq a)^L\}$ as the set Y in the above and get $\{y \mid (y \subseteq a)^L\} \subseteq L_\alpha$, i.e. $\forall y \in L(y \subseteq a \Rightarrow y \in L_\alpha)$ and so $\exists x \in L \forall y \in L(y \subseteq a \Rightarrow y \in x)$ which is the relativization required.

For the weak replacement axiom, suppose that $\psi(z,y)$ is a formula which when relativized to L and restricted to L gives a partial function; this is the same as $(\forall z, u, v(\psi(z,u) \wedge \psi(z,v) \Rightarrow u = v))^L$. Then if $a \in L$, the image of a under ψ^L will be a set, by the replacement axiom; take this image as Y in the above and suppose $Y \subseteq L_\alpha$. Then L_α will satisfy $(\forall y (\exists z \in a \, \psi(z,y))^L \Rightarrow y \in L_\alpha$. So the weak replacement axiom, relativized to L, will hold. □

Now we must show that the relativized subset axioms hold, and the reflection principle will be used for this.

Theorem 8.4.4 *L satisfies the subset axioms.*

Proof We must show that for each formula $\varphi(x, y_1, \ldots, y_k)$ and each $a, b_1, \ldots, b_k \in L$, the set $Y = \{x \in L \mid x \in a \wedge \varphi^L(x, b_1, \ldots, b_k)\}$ is in L. To do this we first assume that φ is written in a form without abstraction terms, and then reflect the formula φ in the L hierarchy. By 8.3.1 we get an ordinal β with $\beta > \text{od}(a), \text{od}(b_1), \ldots, \text{od}(b_k)$ such that

$$\forall x, y_1, \ldots, y_k \in L_\beta(\varphi^{L_\beta}(x, y_1, \ldots, y_k) \Leftrightarrow \varphi^L(x, y_1, \ldots, y_k)).$$

This means that

$$Y = \{x \in L \mid x \in a \wedge \varphi^L(x, b_1, \ldots, b_k)\} = \{x \in L_\beta \mid x \in a \wedge \varphi^{L_\beta}(x, b_1, \ldots, b_k)\}.$$

But then Y is defined in L_β by the formula $x \in \overset{\circ}{a} \wedge \varphi(x, \overset{\circ}{b_1}, \ldots, \overset{\circ}{b_k})$ and so is in $L_{\beta+1}$. □

Before we go on to prove that L is in fact a model of the axiom of choice and the generalized continuum hypothesis, we show that the definition of L itself

CONSTRUCTIBLE SETS AND FORCING

is absolute, at least for L itself. If we write $V = L$ for the formula $\forall x(x \in L)$, i.e. $\forall x \exists \alpha(x \in L_\alpha)$, (as is commonly done, even though we may not be in a language which allows these class constants), what we shall do next amounts to showing that $V = L$ is absolute for L. This will be $\forall x \in L \exists \alpha \in L(x \in L_\alpha^L)$ and will follow when we note that every ordinal is in L (since $\alpha \in L_{\alpha+1}$ as noted in the proof of 8.4.1), and use the next result:

Lemma 8.4.5 *For each α, $L_\alpha^L = L_\alpha$.*

Proof What we shall do is first to show that $\mathcal{D}(X)$ is absolute for L_λ for limit $\lambda > \omega$. Then we go back to the proof of transfinite recursion in 6.6.4, and show, for $\alpha \geq \omega$, that the **good** sequences which were used in that proof, in the case of the actual definition of L_α, will themselves be in L_λ for limit $\lambda > \alpha$, and hence show that there is a term for L_α which is absolute between L_λ and V, for any limit $\lambda > \alpha$. We shall make repeated use of the criteria for formulas and terms to be absolute given in 8.2.6.

First, for $\alpha \geq \omega$, we note that all the Gödel-sets of variables or constant terms of $\mathcal{L}_{L_\alpha}$, i.e. the sets $\ulcorner v_i \urcorner$ and $\ulcorner \overset{\circ}{x} \urcorner$ for $x \in L_\alpha$, will be ordered pairs which are members of $L_{\alpha+2}$. This just uses the fact that if $a, b \in L_\alpha$ then $\{a, b\} \in L_{\alpha+1}$ and $\langle a, b \rangle \in L_{\alpha+2}$, for any α.

The formulas of the language can build to any finite height above these, but they are all built from these sets $\ulcorner v_i \urcorner$ and $\ulcorner \overset{\circ}{x} \urcorner$ for $x \in L_\alpha$, together with members of ω, using finite sets of ordered pairs. The same will be true for the sequences of subformulas x which were defined in 8.1.2 by $\mathrm{FmlSeq}(u, x, n, X)$. So for the case $X = L_\alpha$ or $X \in L_\alpha$, if $\mathrm{FmlSeq}(u, x, n, X)$ holds, then $x \in L_{\alpha+\omega}$ and $\mathrm{FmlSeq}(u, x, n, X)$ and $\mathrm{Fml}(u, x)$ will be absolute for $L_{\alpha+\omega}$.

Similarly if $X = L_\alpha$ or $X \in L_\alpha$, the sequences used as assignment sequences in defining the satisfaction sequences over X are members of $^r X$ and will be in $L_{\alpha+\omega}$, and for each value of a satisfaction sequence, $S(i)$ will be $\{a \in {}^r X \mid (X, \in) \models \varphi(a)\}$ for some φ. We must show that these values are in $L_{\alpha+\omega}$ for all formulas φ of $\mathcal{L}_X$, and it will then follow that the satisfaction sequences themselves (since they are finite sequences) will be in $L_{\alpha+\omega}$. This will be just what is needed to apply 8.2.6(ii) to show that $\mathcal{D}(X)$ is absolute for $L_{\alpha+\omega}$.

We show this by induction on the construction of the formula, i.e. if S is a satisfaction sequence, by induction on its length. If the length is one, we have an atomic formula, and $S(0)$ is $\{a \in {}^r X \mid a(j) \in a(l)\}$ or one of seven other possibilities as in 8.1.3.

First note that if $a \in {}^r X$ and $u \in a$ then $u = \langle i, y \rangle$ for $i \in \omega$ and $y \in X$, so $u \in L_{\alpha+2}$ (remember, $X = L_\alpha$ or $X \in L_\alpha$). So since a is finite, $a \in L_{\alpha+3}$. So $^r X$ is defined over $L_{\alpha+3}$ by $\{a \mid \mathrm{Func}(a) \wedge \mathrm{dom}(a) = r \wedge \forall i < r(a(i) \in \overset{\circ}{X}\}$, which is absolute by 8.2.6(ii) and so $^r X \in L_{\alpha+4}$.

This immediately gives one of the cases wanted, and another is the empty set which is easy. The other cases now all follow in the same way: they are defined over $L_{\alpha+4}$ by formulas such as $\{a \in \overset{\circ}{Y} \mid a(i) \in a(l)\}$ or $\{a \in \overset{\circ}{Y} \mid \overset{\circ}{z} = a(j)\}$, etc., where $Y = {}^r X$. These are absolute, and so all satisfaction sequences over $X = L_\alpha$ will have $S(0) \in L_{\alpha+5}$.

Now the induction steps for negations and implications are immediate, since if $Y, Z \in L_\beta$ then $Y - Z$ and $Y \cup Z$ are in $L_{\beta+1}$, defined by $\{v \mid v \in \overset{\circ}{Y} \wedge v \notin \overset{\circ}{Z}\}$ and $\{v \mid v \in \overset{\circ}{Y} \wedge v \in \overset{\circ}{Z}\}$. And for existential quantifications, if $x(i) = \ulcorner \exists v_j x(l) \urcorner$ (where $\operatorname{SatSeq}(S, x, X)$) then $S(i)$ must be

$$\{a \in {}^r X \mid \exists b \in S(l)[\forall m < r(m \neq j \Rightarrow a(m) = b(m))]\};$$

and this again is absolute, and shows that if $S(l) \in L_\beta$ then $S(i) \in L_{\beta+1}$.

To complete the picture, it is easy to see that $\operatorname{proj}_i(x)$ is absolute, and we have all the ingredients to apply 8.2.6(ii) to the definition 8.1.4 of $\mathcal{D}(X)$ for the case $X = L_\alpha$ or $X \in L_\alpha$. We have shown:

Lemma 8.4.6 $\mathcal{D}(\overset{\circ}{X})$ *is absolute when* $X = L_\alpha$ *or* $X \in L_\alpha$ *for* $\alpha \geq \omega$, *for any* L_λ *with limit* $\lambda > \alpha$. *Hence* $\mathcal{D}(X)$ *is absolute for* L_λ. $\square$

Now for the transfinite recursion. It is not obvious what will be absolute for finite L_n, nor is it really of interest, so we start with $L_\omega = V_\omega$ which is absolute. We can then define:

Definition 8.4.7 L-**good** f for

$$\operatorname{Func} f \wedge \operatorname{Ord}(\operatorname{dom} f) \wedge (\omega \in \operatorname{dom} f \Rightarrow f(\omega) = L_\omega) \wedge \forall \beta \in \operatorname{dom} f$$

$$[[\beta < \omega \Rightarrow f(\beta) = \emptyset] \wedge [(\operatorname{Lim} \beta \wedge \beta > \omega) \Rightarrow f(\beta) = \bigcup_{\alpha < \beta} f(\alpha)] \wedge$$

$$\wedge [\beta + 1 \in \operatorname{dom} f \Rightarrow f(\beta + 1) = \mathcal{D}(f(\beta))]].$$

So these L-**good** functions are sequences of some ordinal length, which give $\emptyset$ until ω but thereafter will satisfy $f(\alpha) = L_\alpha$ as long as they are defined. All parts have been shown absolute, for limit L_λ, and so the whole is absolute for these L_λ also.

Further, we need to show that if $\beta < \lambda$ then there will be an L-**good** function in L_λ with domain β (in fact it will be unique). We prove this by induction on β. For $\beta \leq \omega$ it is an easy exercise; and for successors the induction step is simple: given $f \in L_\lambda$ with domain β where $\beta = \gamma + 1$, let $f' = f \cup \langle \gamma, \mathcal{D}(f(\gamma)) \rangle$. We shall have $f(\gamma) \in L_\alpha$ for some $\alpha < \lambda$, since λ is a limit, and so $\mathcal{D}(f(\gamma)) \in L_\lambda$ and $f' \in L_\lambda$.

For β a limit, using the induction hypothesis, we will have for $\omega \leq \alpha < \beta$ that

$$v = L_\alpha \Leftrightarrow \exists f(L\text{-}\mathbf{good}\, f \wedge \alpha \in \operatorname{dom} f \wedge v = f(\alpha))$$

CONSTRUCTIBLE SETS AND FORCING

and there will be such an L-**good** f with $f \in L_\beta$. So $v = L_\alpha$ is absolute for L_β and the term

$$\{\langle \alpha, v\rangle \mid \exists f(L\text{-}\mathbf{good}\, f \wedge \alpha \in \operatorname{dom} f \wedge v = f(\alpha))\}$$

will define absolutely the L-**good** sequence of length β within L_β. So that sequence will be in $L_{\beta+1}$, and hence in L_λ as required.

8.4.8 Exercise

(1) Show that all these results will hold for L itself as well as for L_λ for limit $\lambda > \omega$.

We have now completed the proof that L_α is absolute between L_λ and L, and between L and V, if we make use of 8.2.5(1). So, as noted earlier, we can now say that $V = L$ holds in L, and using the results up to 8.4.4 we have shown that L is a model of the axioms of ZF together with $V = L$. This is our first example of what is called an *inner model*. In contrast with the notions discussed in chapter 2, where we took what in effect are external notions to the language such as the cumulative type structure, and made models from those, here we are taking relativized formulas of the language as providing our model, in the sense that we have proofs of the relativized axioms. From a formal point of view, one may be reluctant to accept the external notions; but the inner models present no such problems. However it is also clear that inner models cannot hope to prove the consistency of the original system, since their properties are proved within that system (which would be able to prove anything if it were inconsistent). So inner models can only provide relative consistency results: so far we have shown that if ZF is consistent, so is ZF plus $V = L$.

8.5 The axiom of choice in L

The idea of why the axiom of choice should hold in L is fairly straightforward, but, as in the last section, when we come to details, we find that we must carefully check absoluteness. First we sketch the idea.

We shall show that we can define a well-ordering of each L_α in a uniform way, which will carry through into a definable well-ordering of the whole of L. Given a well-ordering of the whole of L, we shall get a well-ordering of any subset of L, and hence within L every set will have a well-ordering and the axiom of choice will hold.

In defining the well-ordering of L, we decide first that sets with lower constructible order will come first. So if the order which we are defining is to be written as $y <_L z$, for $y, z \in L$, we will want to ensure $\operatorname{od} y < \operatorname{od} z \Rightarrow y <_L z$.

This will have one important consequence immediately: the predecessors of any element under $<_L$ will always be a set, and not a proper class. Since we are well-ordering the whole proper class L, this is not automatic from the start. But if $y \in L_\alpha$ then any predecessor under $<_L$ will also be in L_α and L_α is a set.

Next we have to decide how to well-order sets with the same constructible order. For this, we note how they arise: each such y is given as some projection $y = \text{proj}_i S(j)$, where S is a satisfaction sequence for a formula of $\mathcal{L}_X$, with $X = L_\alpha$ and $\alpha + 1 = \text{od}\, y$. The satisfaction sequences depend only on the formula sequence x for which $\text{SatSeq}(S, x, X)$ holds, since X is not varying here. So it is sufficient to well-order the formula sequences (for formulas of $\mathcal{L}_X$) and we shall be able to get from that our desired well-ordering of $L_{\alpha+1} - L_\alpha$.

We give one way of completing these details:

Definition 8.5.1 (i) $y = \text{spec}(\alpha, x, i)$ for

$$\exists u, n, S(\text{FmlSeq}(u, x, n, L_\alpha) \wedge \text{SatSeq}(S, x, L_\alpha) \wedge y = \text{proj}_i(S(n))).$$

(x is a formula sequence which gives y as a projection of the last member of its satisfaction sequence in L_α.)

We have eliminated one of the redundancies of definition 8.1.4 here in assuming that y is given as a projection of the *last* member of S and not just *some* member. This is safe, since if we truncate any formula sequence we shall get another formula sequence, and we want to cut out some of the redundancy here.

Now the well-ordering of $L_{\alpha+1}$ will take the form, for $y, z \in L_{\alpha+1}$:

(ii) $y <_{L_{\alpha+1}} z$ for

$$\text{od}\, y < \text{od}\, z \vee [\text{od}\, y = \text{od}\, z = \beta < \alpha + 1 \wedge y <_{L_\beta} z] \vee$$
$$\vee \exists x, i \forall x', i'[y = \text{spec}(\alpha, x, i) \wedge [z = \text{spec}(\alpha, x', i') \Rightarrow$$
$$\Rightarrow (x <^* x') \vee (x = x' \wedge i < i')]],$$

where $x <^* x'$ must still be defined, this will be the desired well-ordering of the formula sequences. Finally we shall want:

(iii) $y <_{L_\lambda} z$ where λ is a limit ordinal and $y, z \in L_\lambda$, for

$$\exists \alpha < \lambda (y, z \in L_{\alpha+1} \wedge y <_{L_{\alpha+1}} z),$$

and the corresponding definition:

(iv) $y <_L z$, for $y, z \in L$, for

$$\exists \alpha (y, z \in L_\alpha \wedge y <_{L_\alpha} z).$$

It is easier to describe $x <^* x'$, the well-ordering for formula sequences, than to write the formula. First, we shall say that shorter formula sequences

CONSTRUCTIBLE SETS AND FORCING

precede longer ones, i.e. we use the clause

$$\operatorname{dom} x < \operatorname{dom} x' \Rightarrow x <^* x'.$$

Next, for formula sequences of the same length, we find the first difference, i.e. $k < \operatorname{dom} x = \operatorname{dom} x'$ for which $x(k) \neq x'(k)$ but $x(j) = x'(j)$ for all $j < k$. Then we compare $x(k)$ and $x'(k)$. Each will satisfy one of Atfml, Nfml, Ifml, and Efml (from 8.1.2) and we decide to take them in this order, so we shall have the clause

$$(\operatorname{Atfml}(x(k), L_\alpha) \wedge \operatorname{Nfml}(x'(k), x'(j))) \Rightarrow x <^* x',$$

and five other similar clauses to give effect to this.

Finally we must order x and x' when $x(k)$ and $x'(k)$ are of the same type. If both are negations, we must have $\operatorname{Nfml}(x(k), x(j))$ and $\operatorname{Nfml}(x'(k), x'(j'))$ for $j \neq j'$ since $x(j) = x'(j)$ for all $j < k$. So we can use the order of j and j', and a clause

$$(\operatorname{Nfml}(x(k), x(j)) \wedge \operatorname{Nfml}(x'(k), x'(j'))) \wedge j < j') \Rightarrow x <^* x'.$$

We can treat implications and quantifications similarly, using the first difference between the subformulas or the earlier of the variables being quantified; we leave it as an exercise to write suitable clauses.

For atomic formulas, each has three components, and we can well-order them by taking the first difference of these components, and putting variables $\ulcorner v_i \urcorner$ in the order of the suffix i and before all constant terms $\ulcorner \overset{\circ}{z} \urcorner$, and then taking these constant terms in the order given by $<_{L_\alpha}$ since each of these must have $z \in L_\alpha$. So a typical clause might be

$$x(k)(0) = \ulcorner \overset{\circ}{z}_1 \urcorner \wedge x'(k)(0) = \ulcorner \overset{\circ}{z}_2 \urcorner \wedge z_1 <_{L_\alpha} z_2 \Rightarrow x <^* x'.$$

Note that this is the one place where the definition has to be by recursion: we must use the ordering already defined, $<_{L_\alpha}$, to order the constant symbols that appear in the formulas of $\mathcal{L}_{L_\alpha}$.

We also have to decide which comes first out of $=$ and $\in$; the coding in 8.1.2 suggests that $=$ be first, but that is of course just an arbitrary choice.

8.5.2 Exercises

(1) Complete the definition in 8.5.1 by writing out more of the formula to define $x <^* x'$ for formula sequences; and show that the result is in fact a well-ordering, and hence that $<_{L_{\alpha+1}}$ is a well-ordering.

   [There is no great virtue in writing every clause, but one more of each type would be a useful exercise. The methods sketched in 5.4.6, particularly (3), are needed to show this is a well-ordering. Note that for

limit ordinals and for L itself we are putting together well-orderings which are themselves well-ordered by initial segments, and we need the method of 7.3.4(1) to show that the result is a well-ordering.]
(2) Show that the well-ordering $<_{L_\alpha}$ is absolute for L_λ for any limit $\lambda > \alpha$, and hence also for L.
[This involves just a great deal more of the same sort of work as in 8.4.]

8.6   The generalized continuum hypothesis in L

We shall give a little more than the simplest proof here, since we can introduce some important ideas from model theory which have many uses in set theory. But we sketch first the basic idea of the proof.

We shall show that if $X \in L$ and $X \subseteq \kappa$ where κ is an initial ordinal in L, then the constructible order of X, od X, has $|\text{od } X| \leq \kappa$. (This will be true in L, i.e. all the cardinals will be taken relative to L. Cardinals are not absolute, so it is important to note where we are working; in fact we could restrict ourselves for the whole of this section to working in L.)

This will suffice for the generalized continuum hypothesis, for we can now remember the proof of the power-set axiom in L. We know that $\mathcal{P}^L(\kappa)$ is just the set of all $X \in L$ for which $X \subseteq \kappa$, since $X \subseteq \kappa$ is absolute, and that this is a set because it is a subset of L_α for some α. Using the result above, we see that this α can be taken as the first ordinal for which $|\alpha| \geq \kappa$ in L, since this will be larger than all the constructible orders of subsets of κ in L, and that is all we require of it. This ordinal is usually called the cardinal successor of κ and written κ^+, and we have shown that $\mathcal{P}(\kappa) \subseteq L_{\kappa^+}$, all relativized to L. Now the generalized continuum hypothesis follows, because we proved earlier in 8.2.2 that $|L_{\kappa^+}| = \kappa^+$. Hence we have $|\mathcal{P}(\kappa)| \leq \kappa^+$ in L, for all infinite cardinals κ, and that is the generalized continuum hypothesis.

8.6.1   Skolem functions

First we introduce the idea of Skolem functions. We define them only for the simple structures we are interested in, which are simply structures of the form $(A, \in)$ for some set A. We shall write formulas of the language $\mathcal{L}_A$ with just the variables v_1 free as $\varphi(v_1, \vec{\overset{\circ}{y}})$, where $\vec{\overset{\circ}{y}} = \overset{\circ}{y}_1, \ldots, \overset{\circ}{y}_k$ is a list of all the constant symbols appearing in φ; and write $\varphi(v_1, \vec{v})$ for the formula with all the constant symbols replaced by variables other than v_1. Suppose that $f: A^k \to A$ is a function such that, given $\vec{y} = y_1, \ldots, y_k \in A$, we have that whenever $(A, \in) \models \exists v_1 \varphi(v_1, \vec{\overset{\circ}{y}})$, then $f(\vec{y}) = z$ for some $z \in A$ for which $(A, \in) \models \varphi(\overset{\circ}{z}, \vec{\overset{\circ}{y}})$; then f is called a *Skolem function* for φ in the structure

CONSTRUCTIBLE SETS AND FORCING   149

$(A, \in)$.

The point of such functions is that when we want to check absoluteness for the formula $\exists v_1 \varphi(v_1, \vec{v})$, using the criterion in 8.2.6(i), the values taken by the Skolem function will be suitable ones to use, as in the next lemma. In general we want to consider sets of Skolem functions, and we shall say that F is a *set of Skolem functions for the structure* $(A, \in)$ if F contains one Skolem function for each formula of the language $\mathcal{L}$ with v_1 free (but without constant symbols). Note that such a set will be countable, since we are considering only formulas of the language without constant terms, and there are only countably many of them. With this definition we can show:

Lemma 8.6.2 *Let* $Y \subseteq A$ *be such that, for some set* F *of Skolem functions for the structure* $(A, \in)$, *the set* Y *is closed under each* $f \in F$. *Then every formula of the language* $\mathcal{L}_Y$ *is absolute between* Y *and* A.

Proof By induction on the formulas without abstraction terms (using as usual that every formula has an equivalent formula without abstraction terms). Atomic formulas are always absolute, and from what we noted in 8.2.5(2), it follows that absoluteness can only fail if we have a formula of the form $\varphi(v_1, \vec{v})$ which is absolute, and some assignment of $\overset{\circ}{\vec{v}}$ in the smaller structure (in this case in Y), say $\overset{\circ}{\vec{y}}$, such that $\exists v_1 \varphi(v_1, \overset{\circ}{\vec{y}})$ holds in A but fails in Y. But then if $f_\varphi \in F$ is a Skolem function for φ in A, we cannot have $f_\varphi(\overset{\circ}{\vec{y}}) \in Y$ (otherwise $\exists v_1 \varphi(v_1, \overset{\circ}{\vec{y}})$ would hold in Y also). But this contradicts the assumption that Y is closed under f_φ. □

Corollary 8.6.3 *(AC) Given any* $Y \subseteq A$, *we can find* B *with* $Y \subseteq B \subseteq A$ *and* $|B| \leq |Y| + \aleph_0$, *such that every formula is absolute between* B *and* A.

Proof Take a set F of Skolem functions for A and let B be the closure of Y under the functions of F. The last lemma shows that everything will be absolute between B and A. We shall need the axiom of choice in general to show the existence of the set F, and to compute the cardinal of B. The closure of Y under F can be built up in ω stages Y_n, with $Y_0 = Y$ and at each stage applying each member of F to every member of Y_n^r for the appropriate r, to form Y_{n+1}. B will be $\bigcup_{n < \omega} Y_n$, and we shall have $|Y_n| = |Y_{n+1}|$ at every stage, since F is countable, except for the first in the case Y is finite, when we will get $|Y_1|$ is countable (it could be finite for very uninteresting A). □

The substructure $(B, \in)$, constructed as in this proof by closing a given set Y under a set of Skolem functions F, is called the *Skolem hull* of Y under F in $(A, \in)$.

When every formula is absolute between two structures, as in this corollary, the smaller is said to be an *elementary substructure* of the larger (and the larger an *elementary extension* of the smaller). This is a fundamental notion for model theory, and is much used in set theory also; but we shall just make one use of it.

In order to prove the GCH, we have to show, if $X \subseteq \kappa$ is in L, that $|\text{od } X| \leq \kappa$. What we shall do is to take some limit ordinal λ for which $X \in L_\lambda$, and form an elementary substructure B of L_λ which contains as members the set X, and also all ordinals $\alpha \leq \kappa$. This gives a starting set Y of cardinal κ, and so by the corollary we get B also of cardinal κ. Now the set B will in general not be a transitive set; although the Skolem functions will put members into B for any set in B which does have members in L_λ, there is no guarantee that they will put *all* of them in. (In general if λ has cardinal greater than κ there will be members of B of cardinal greater than κ, and B must omit most of their members.)

There are two further steps now. First we take this set B and find an isomorphic set which *is* transitive. This step is by forming what is known as the *transitive collapse* of B. This can be defined very simply: we define by induction on the $\in$ relation, a function π (the *collapsing map*), defined by:

Definition 8.6.4 For any set B, the transitive collapse of B is the map π given by
$$\pi(z) = \{\pi(y) \mid y \in z \cap B\}$$
for $z \in B$.

This can be thought of as a definition by induction on the rank of z. Let $Z = \pi(B)$ be the image of this map. Then the second step is to show that this set Z must be of the form L_β for some β. We shall see that the original set X will be in Z. The requirement that $\{\alpha \mid \alpha \leq \kappa\} \subseteq Y$ will ensure that each $\alpha \leq \kappa$ will have $\pi(\alpha) = \alpha$, and then $\pi(X) = X$ will follow from the definition of π. But this will complete the proof of the GCH, since we have $|L_\beta| = |\beta|$ and $|L_\beta| = |B| = \kappa$ and $X \in L_\beta$.

So we need:

Lemma 8.6.5 *The map π is an isomorphism between B and its image $Z = \pi(B)$, and this image Z is transitive. Also $\pi(X) = X$.*

Proof That Z is transitive is immediate from the definition; that π is one-to-one requires that B satisfies the axiom of extensionality. This will hold in B because it holds in L_λ, and B is an elementary substructure of L_λ.

So if $y \neq y'$ are two members of B, then some $z \in B$ must be in $y - y'$ or in $y' - y$. So $\pi(z) \in \pi(y) - \pi(y')$ or $\pi(z) \in \pi(y') - \pi(y)$, and $\pi(y) \neq \pi(y')$.

That π preserves $\in$ is immediate: if $z \in y$ and z and y are both members of B then $\pi(z) \in \pi(y)$ (this is in effect the definition of π).

CONSTRUCTIBLE SETS AND FORCING   151

Now we note that if $z \in B$ is a transitive set with also $z \subseteq B$, then $\pi(z) = z$, and indeed $\pi(u) = u$ for every member $u \in z$. This follows by induction on the definition of π, since $z \subseteq B$ implies that for each of these $u \in z$ we shall also have $u \subseteq B$, since $u \subseteq z$. So the definition of π gives

$$\pi(u) = \{\pi(w) \mid w \in u \cap B\}$$

but $u \cap B = u$ and the induction hypothesis is $\pi(w) = w$ for each $w \in u$; hence $\pi(u) = \{w \mid w \in u\} = u$, and similarly $\pi(z) = z$.

We ensured from the start that $\kappa \subseteq B$, so applying this to each $\alpha \leq \kappa$ shows that $\pi(\alpha) = \alpha$. Then we have $\pi(X) = \{\pi(\alpha) \mid \alpha \in X\} = X$ since also $X \subseteq \kappa \subseteq B$. □

The final result we need is now:

Theorem 8.6.6 *If Z is the transitive collapse of an elementary substructure of L_λ for limit ordinal λ then $Z = L_\beta$ for some limit ordinal β.*

Proof We shall assume that $B \subseteq L_\lambda$ is the elementary substructure, and that $\pi : B \to Z$ is the collapsing isomorphism, as above.

Since Z is transitive, so is $\text{On} \cap Z$ and so this must be an ordinal; let $\text{On} \cap Z = \beta$, and we shall show that $Z = L_\beta$. This β must be a limit ordinal: for suppose $\alpha \in \beta$, then $\alpha \in Z$ so $\alpha = \pi(\gamma)$ for $\gamma \in B$. Now $L_\lambda \models \exists \delta(\overset{\circ}{\gamma} \in \delta)$ so $B \models \exists \delta(\overset{\circ}{\gamma} \in \delta)$ also; say $B \models \overset{\circ}{\gamma} \in \zeta \wedge \text{Ord}\,\overset{\circ}{\zeta}$ for $\zeta \in B$. Apply the collapse π: we get $\pi(\gamma) = \alpha$ and $\pi(\zeta) = \xi$, say, and ξ must be an ordinal in Z with $\alpha \in \xi$, since π is an isomorphism and so $Z \models \overset{\circ}{\alpha} \in \overset{\circ}{\xi} \wedge \text{Ord}\,\overset{\circ}{\xi}$, and this formula is absolute for transitive structures (by 8.2.7). So we have $\alpha \in \xi \in \beta$ and we cannot have $\beta = \alpha + 1$.

To show that $Z = L_\beta$, we use more arguments similar to the above, but the absoluteness requires very much more checking. First, we know that $L_\lambda \models \forall y \exists \alpha (y \in L_\alpha)$, and so B and hence Z satisfies the same formula. The formula $y \in L_\alpha$ is absolute for L_λ by the work in 8.4.5. We need to know that it is absolute for Z also, and this will be the final part of the proof. From it we can deduce $Z \subseteq L_\beta$.

For the other way, we know that $L_\lambda \models \forall \alpha \exists u(u = L_\alpha)$, and so B and hence Z satisfies this formula. Now we need to know that $u = L_\alpha$ is absolute for Z, and we shall be able to deduce $L_\beta \subseteq Z$.

This final part is now very much parallel to the work in 8.4.5. There we had to show that the formula sequences and the satisfaction sequences were in L_λ; and then that the L-**good** functions were in L_λ (providing the constant symbols appearing were in L_λ). Now we shall need to show that they are in Z (providing the constant symbols are in Z). For this, we first note that at each step in the argument showing that formula sequences and satisfaction sequences had to be in L_λ, we actually showed that they were uniquely definable by some formula of the language with appropriate parameters. So when we form the elementary substructure B we must include these sequences; if we did not, some formula asserting the existence and uniqueness of one of these sequences would not be absolute.

Now if these sequences are in B, their images under the collapse π will be in Z, and will satisfy the same definitions in Z (with appropriately mapped parameters) as they did in L_λ. These definitions use bounded quantifiers only, and will be absolute for Z since Z is transitive. Hence Z will have all the formula sequences and satisfaction sequences needed to prove that $\mathcal{D}(X)$ is absolute for Z.

Finally, the L-**good** functions required to show that the recursive definition of L_α is absolute for L_λ, are also uniquely definable in L_λ and so must be in B, and their images must be in Z. Again their definitions will be absolute for Z, using the absoluteness of $\mathcal{D}(X)$, and so these images, which will satisfy these definitions in Z, must in fact be L-**good** sequences, now of the length which is the image under π of their original length. These image lengths will be all the ordinals of Z, since the original lengths were all the ordinals of L_λ, and this suffices to show that $u = L_\alpha$ is absolute for Z as required. □

8.6.7   Exercise

(1) In the definition of Skolem functions we did not make any restriction concerning the value to be taken by the Skolem function when there was no element satisfying the appropriate formula; as we have defined them, the Skolem functions can take arbitrary values in these cases. Show that this does not affect the proof of 8.6.3.

Show that if we restrict these extra values (say by taking ∅ as the value whenever no element satisfies the appropriate formula), then we can find a set F of Skolem functions for L_λ for limit λ such that every member of F is definable in L_λ.

[Use the definable ordering $<_{L_\lambda}$ and take the first element under this ordering which satisfies the appropriate formula.]

Hence show the following strengthening of 8.6.3: given $Y \subseteq L_\lambda$ there will be an elementary substructure B of L_λ containing Y which is minimal in the sense that it will be an elementary substructure of every elementary substructure of L_λ containing Y.

[This B will result from using the definable Skolem functions. This B is known as the *definable Skolem hull* of Y in L_λ.]

CONSTRUCTIBLE SETS AND FORCING

8.7 Another presentation

An alternative presentation of the constructible sets is given by Gödel in [Göd40], where instead of taking the definable power-set operation in one go, he uses just eight relatively simple functions, which are all iterated in turn, together with the operation of collecting together all the sets obtained so far into a set. He shows that these will give very much the same results: all the sets which result, will turn up in some definable power-set, and all the members of the definable power sets will in fact turn up at some stage as the value given by one of the functions. The functions used are based on work of Bernays, who showed that the subsets defined by formulas can be given by a finite set of functions. The eight functions are the following:

$$F_1(x,y) = \{x,y\}$$
$$F_2(x,y) = \{\langle u,v\rangle \mid \langle u,v\rangle \in x \land u \in v\}$$
$$F_3(x,y) = x \setminus y$$
$$F_4(x,y) = x \restriction y$$
$$F_5(x,y) = x \cap \mathrm{dom}\, y$$
$$F_6(x,y) = x \cap y^{\cup}$$
$$F_7(x,y) = \{\langle u,v,w\rangle \in x \mid \langle v,w,u\rangle \in y\}$$
$$F_8(x,y) = \{\langle u,v,w\rangle \in x \mid \langle u,w,v\rangle \in y\}$$

8.7.1 Exercises

(1) Show that L_λ is closed under each of the functions $F_1,\ldots,F_8$ for limit λ. Show also that any set in $\mathcal{D}(X)$ can be given by a finite sequence of the operations $F_1,\ldots,F_8$ starting from members of X.

[It is easiest to show first that the satisfaction sets themselves, rather than the projections of these as were used to give $\mathcal{D}(X)$, are all of them given by such finite sequences of these operations. Then the projections required can be got by further applications. Many other sets of functions have been given with the same properties, see e.g. Jech [Jec73].]

(2) A transitive class M is called *almost universal* if every subset of M is included in an element of M:

$$\forall s \subset M \exists y \in M(s \subseteq y).$$

(In 8.4.3 we showed that L is almost universal.)

Show that if a class M is transitive and almost universal, and is closed under the eight Gödel functions, then M is a model of ZF.

[Closure under the eight functions is used to show that the subset axioms hold; the power-set and replacement axioms in their weak forms follow from the almost universality as in 8.4.3. Further details of this sort of presentation of L are in [Jec73].]

We shall not give further details of this sort of presentation here. Yet another presentation has been given by Jensen [Jen72] in his work on the fine structure of L.

8.8 Forcing models

These were introduced following Paul Cohen's proof of the independence of the GCH and the axiom of choice. Cohen's original description followed Gödel's work fairly closely, and was presented in terms of adding new subsets to L (subsets which were not constructible in Gödel's sense). Later presentations generalized this to adding new subsets to any model, and then introduced *Boolean-valued models* to present equivalent ideas. We present the version which allows us to add new subsets to any model; it owes a lot to work of Shoenfield ([Sho71]). Then we sketch the Boolean-valued models.

8.8.1 Forcing conditions

The basic discovery of Cohen [Coh63a] was a method of adding new sets to a model of ZFC to get another, larger model of ZFC. To present this method in some generality, but in as simple a way as we can, we shall follow Cohen in assuming that we are starting with a model M (really $\langle M, \in \rangle$) of ZFC which is a transitive, $\in$-model, and which is *countable*. (Since our models will always be $\in$-models, we shall write just M instead of $\langle M, \in \rangle$ throughout this section.)

This is a stronger assumption than just the assumption that ZFC is consistent, and later, in 8.14, we shall indicate how it could be avoided. But it seems a natural assumption, indeed it seems almost as natural as the assumption of consistency, when we remember that the justification we gave for the axioms of ZFC was that they should describe the cumulative hierarchy of sets, and that we then want to add axioms of extent to say that this hierarchy has no conceivable end. All the further axioms of extent that we consider do in fact imply the existence of countable, transitive $\in$-models of ZFC.

The word *forcing*, which is usually used to describe this work, goes back to the earliest presentations, and can be explained as follows. We start with a partial order $\langle P, \preccurlyeq \rangle$ in the model (the *ground* model) M of ZFC. We seek to add a *generic subset* G of P to the model to form a new model, the *generic extension* $M[G]$ which extends the ground model and is still a model of ZFC. (We shall have $M \subseteq M[G]$ and $G \in M[G]$.)

If we start with M countable, this will be possible: for any poset P in M (we

CONSTRUCTIBLE SETS AND FORCING

shall, as here, usually write P when we should write $\langle P, \preccurlyeq \rangle$), we shall prove that there is a subset G generic over M. Provided P is not too simple, it will be easy to see that such a generic subset must be new, i.e. not in M already; but the important point is that its properties will be decided within M. The basic information one has about G will be of the form: $p \in G$ for specific $p \in P$ (such p are called *forcing conditions*, or just *conditions*). G must be consistent, in that any two members of G must be *compatible*, where $p, q \in P$ are compatible if they have a common extension $r \in P$. So M can deduce from $p \in G$ that every member of G must be compatible with p; and the condition p is then said to *force* this extra information. Further information will follow from the fact that G is generic, but nothing more than this will be forced by p. All this information will be given in an appropriate forcing language, and we shall show that the *forcing relation* p forces φ (denoted by $p \Vdash \varphi$, where φ is a sentence of the appropriate forcing language), will be definable in M. This will be the main tool in showing that the resulting model $M[G]$ is in fact a model of ZFC, and in deciding its other properties, such as whether it satisfies the continuum hypothesis.

8.8.2 Warning

When reading any work on forcing, it needs to be noted that there is a confusion about the partial orderings used, as to which way round they should be taken: if p extends q, is $p \preccurlyeq q$ or $q \preccurlyeq p$? Both can be made to seem reasonable in different contexts; the basic idea is that an extension should give *more* information. As a forcing condition is extended, the collection of all information forced gets *larger*; but the collection of possible generic subsets gets *smaller*. Most (but not all) recent work seems to accept this last point of view, and say that p extends q if $p \preccurlyeq q$, and we shall follow that view here (it fits best with the Boolean-valued models). But in reading any work on forcing the reader needs to be alert, and always check which convention is being used. (It is usually obvious from the working, if not stated at the outset.)

Definition 8.8.3 We assume (as above) that M is a countable transitive $\in$-model of ZFC, and that $\langle P, \preccurlyeq \rangle$ is a partial ordering with $\langle P, \preccurlyeq \rangle \in M$, and $P \neq \emptyset$.

(i) $p \| q$ (for $p, q \in P$) for $\exists r \in P (r \preccurlyeq p \land r \preccurlyeq q)$
   (p and q are *compatible*).
(ii) $p \perp q$ (for $p, q \in P$) for $\forall r \in P \neg (r \preccurlyeq p \land r \preccurlyeq q)$
   (p and q are *incompatible*).
   For $D \subseteq P$,
(iii) D is *dense* if $\forall p \in P \exists d \in D (d \preccurlyeq p)$.
   And for $G \subseteq P$, G is *P-generic* over M if

(iv) $\forall p, q \in G \exists r \in G(r \not\preceq p \wedge r \not\preceq q)$ (G is *consistent*);
(v) $\forall p \in G \forall q \in P(p \not\preceq q \Rightarrow q \in G)$ (G is *closed upward*, i.e. weaker conditions are in if stronger ones are); and
(vi) $\forall D \subseteq P(D$ is dense and $D \in M$ implies $G \cap D \neq \emptyset)$ (G meets every dense set in M).

We can already see why we shall want M to be countable:

Lemma 8.8.4 *If M is countable, then for any partial ordering $P \in M$ and any $p \in P$ there will be a subset $G \subseteq P$ with $p \in G$ which is P-generic over M. Moreover, if P satisfies $\forall p \in P \exists q, r \in P(q \not\preceq p \wedge r \not\preceq p \wedge r \perp q)$, then we shall have $G \notin M$.*

Proof Since M is countable, there can only be countably many dense $D \subseteq P$ with $D \in M$; let them be $(D_n)_{n \in \omega}$. Now define a sequence $(p_n)_{n \in \omega} \subseteq P$ recursively, starting with $p_0 \in D_0$ with $p_0 \not\preceq p$ and so that:

$$p_n \in D_n \wedge \forall i < n (p_n \not\preceq p_i).$$

The fact that D_n is dense ensures that we can find p_n, and we define G by

$$p \in G \Leftrightarrow \exists n (p_n \not\preceq p) \text{ for } p \in P.$$

To see that G is consistent, suppose $q, q' \in G$: say $p_{n_0} \not\preceq q$ and $p_{n_1} \not\preceq q'$. Then either $n_0 \leq n_1$, so that $p_{n_0} \not\preceq q \wedge p_{n_0} \not\preceq q'$, or $n_1 \leq n_0$ so that $p_{n_1} \not\preceq q \wedge p_{n_1} \not\preceq q'$. It is immediate that G is closed upward, and since each $p_n \in G$ it will meet every dense set D which is in M.

To see that $G \notin M$, consider $D_0 = P - G$; clearly this is in M if and only if G is. But D_0 is dense: given any $p \in P$, the hypothesis given (which might be read as: every condition *splits*) says that p can be extended to two incompatible conditions. Since G is consistent, at most one of them can be in G, and the other must be in D_0. And clearly G cannot meet D_0. □

This property of a forcing partial order (that every condition splits) will be called *non-trivial*. There are many other conditions and definitions which are of interest, some of these are in the exercises.

8.8.5 Exercises

(1) For each $p \in P$ let $D_p = \{q \in P \mid q \not\preceq p \vee q \perp p\}$. Show that this will be dense and in M.
  [For any $r \in P$ either $r \perp p$ (so $r \in D_p$), or $r \| p$ in which case there is a common extension q, so we have both $q \not\preceq r$ and $q \in D_p$. This shows that D_p is dense. Show that D_p will be in M since it has a simple definition

using bounded quantifiers from the parameters p, P and $\preccurlyeq$ which are in M, and M satisfies the subset axiom.]

Now let $D_0 = \{p \in P \mid \exists q \in G(p \perp q)\}$. Show that this will also be dense and cannot be in M, if P is non-trivial.

[Given $p \in P$, if $p \notin G$ then no extension of p can be in G (since G is closed upward). So let $q \in G \cap D_p$; we must have $q \perp p$, and so $p \in D_0$ already. If $p \in G$, non-triviality means that we can extend p to incompatible q and r, and at most one of these can be in G. So the other must be in D_0 and D_0 is dense. If $D_0 \in M$ we must have $G \cap D_0 \neq \emptyset$ which contradicts G is consistent.]

(2) A subset $D \subseteq P$ is called *pre-dense* if $\forall p \in P \exists d \in D(p \| d)$. Show we get the same generic sets if we replace (vi) in the definition of generic by: G meets every pre-dense set in M.

[Given a pre-dense set D, let $D' = \{q \in P \mid \exists d \in D(q \preccurlyeq d)\}$ and show D' is dense; and if G is closed upward and meets D' then it meets D.]

(3) A subset $A \subseteq P$ is called an *antichain* in P if its members are pairwise incompatible, i.e. $\forall p, q \in A(p \neq q \Rightarrow p \perp q)$. Show that if A is an antichain in P and $G \subseteq P$ is P-generic over M, then $|A \cap G| \leq 1$, and if A is a *maximal* antichain in P (maximal under $\subseteq$), and $A \in M$, then $A \cap G$ has exactly one member.

[Given a maximal antichain $A \in M$, show that $D = \{p \in P \mid \exists a \in A(p \preccurlyeq a)\}$ is dense and in M.]

8.8.6 The forcing language

We shall construct $M[G]$ using the terms of a language, which we shall take to be identical with the language $\mathcal{L}_M$ introduced in 8.1. But now we shall treat the terms $\mathring{x}$ very differently; in general they will no longer denote x but some set in $M[G]$, which we shall write as x^G (in this context $\mathring{x}$ is often called a *name* or a *term* for x^G). The idea is that we shall really be interested in sets of the form $x = \{\langle p_i, y_i \rangle \mid i \in I\}$ with each $p_i \in P$ and each y_i corresponding to a name $\mathring{y}_i$. (In effect these are relations with domain a subset of P.) Then the term $\mathring{x}$ will denote the set $x^G = \{y_i^G \mid p_i \in G\}$. Note that if $G \notin M$, there is no reason to expect x^G to be in M; but we shall show that there will always be terms which will ensure that $M \subseteq M[G]$ and $G \in M[G]$.

The definition will formally be simpler, in that we shall not in fact demand that we use only terms corresponding to such relations, but actually we shall use *every* term as a name. For any $x \in M$, we shall have the term $\mathring{x}$ in the language $\mathcal{L}_M$, and it will denote:

$$x^G = \{y^G \mid \exists p \in P(\langle p, y \rangle \in x \land p \in G)\}.$$

(Note that we write x^G rather that $\mathring{x}^G$ for this denotation, to keep the notation simpler.) There will be a lot of redundancy, but this will not matter. It will be easier to live with the repetitions that occur than to remove them all; indeed if we tried to remove them all, we may no longer have the important property that the whole construction is, in an important sense, definable in M. Notice that this definition will immediately ensure that the resulting collection will be transitive, since we only put in collections whose members are things already in. Also the induction can be carried out as an induction on the rank (since if $\langle p, y \rangle \in x$ then $\operatorname{rank}(y) < \operatorname{rank}(x)$).

These interpretations will build the extended model $M[G] = \{x^G \mid x \in M\}$, and what will matter most is that we can define the *forcing relation*

$$p \Vdash \varphi(\mathring{x}_1, \ldots, \mathring{x}_n)$$

by a formula of the language $\mathcal{L}_M$ which will be absolute for M, and so that this relation will hold if and only if, for every G which is P-generic over M and for which $p \in G$,

$$M[G] \models \varphi(x_1^G, \ldots, x_n^G)$$

(and in this last satisfaction, we are taking the usual notion of assigning elements x_1^G etc. to appropriate variables, rather than using a language with constant symbols). We want in fact even more; we also will show that whenever

$$M[G] \models \varphi(x_1^G, \ldots, x_n^G),$$

then there is some $p \in G$ such that

$$p \Vdash \varphi(\mathring{x}_1, \ldots, \mathring{x}_n).$$

These facts are summed up in the following lemmas, which we shall prove later: the *truth lemma*,

Lemma 8.8.7 *For any formula $\varphi(\mathring{x}_1, \ldots, \mathring{x}_n)$ of the forcing language $\mathcal{L}_M$ and for any $G \subseteq P$ generic over M, we have*

$$\exists p \in G(p \Vdash \varphi(\mathring{x}_1, \ldots, \mathring{x}_n)) \Leftrightarrow M[G] \models \varphi(x_1^G, \ldots, x_n^G),$$

together with the *definability lemma*,

Lemma 8.8.8 *For any formula $\varphi(\mathring{x}_1, \ldots, \mathring{x}_n)$ of the forcing language $\mathcal{L}_M$ there is a formula $\varphi^*(v_1, \ldots, v_n, v_{n+1}, v_{n+2})$ which is absolute for M and such that*

$$p \Vdash \varphi(\mathring{x}_1, \ldots, \mathring{x}_n) \Leftrightarrow M \models \varphi^*(x_1, \ldots, x_n, p, P).$$

We defer proof of these lemmas until we have shown how all this will be used. First we give the terms needed to show that $M[G]$ does indeed extend M and contain G:

CONSTRUCTIBLE SETS AND FORCING

Definition 8.8.9 For each $x \in M$ we define $\check{x}$ as a term of the forcing language $\mathcal{L}_M$; in this case we shall use $\check{x}$ also for the set of M which gives this term, rather than distinguishing by writing $\overset{\circ}{\check{x}}$ for the term. By recursion we define

$$\check{x} = \{\langle p, \check{y} \rangle \mid y \in x \wedge p \in P\},$$

for $x \in M$. Then $\check{G}$ is the term

$$\check{G} = \{\langle p, \check{p} \rangle \mid p \in P\}.$$

We show these definitions give what we want. For every G which is P-generic over M and every $x \in M$, we shall have $\check{x}^G = x$, or in other words $\check{x}$ denotes x, by induction on rank in M. For the members of $\check{x}^G$ are all z^G for which $\langle p, z \rangle \in \check{x}$ and $p \in G$; by construction $\langle p, \check{y} \rangle \in \check{x}$ for all $y \in x$ and all $p \in P$, and by the induction hypothesis, for each $y \in x$, $\check{y}^G = y$. So $\check{x}$ denotes $\{y \mid y \in x\} = x$ as claimed. So certainly $M \subseteq M[G]$.

Now we can see that $\check{G}$ denotes G since its members will be just the denotations of each $\check{p}$, $\check{p}^G$, for $p \in G$, and each $\check{p}^G = p$. Hence also $G \in M[G]$.

One further fact which follows immediately is that the ordinals of M and of $M[G]$ coincide. This follows by thinking about the rank function, given for any set x by $\text{rank}(x) = \bigcup \{\text{rank}(y) + 1 \mid y \in x\}$, see 6.6.13(3). We can see immediately by induction that $\text{rank}(x^G) \leq \text{rank}(x)$ for any $x \in M$. Each ordinal is the rank of itself, so we have shown that each ordinal of $M[G]$ is a member of an ordinal of M. So we have added no new ordinals in passing from M to $M[G]$. (Note that the rank function is absolute.)

8.9 Forcing in practice: the ZFC axioms hold in $M[G]$

We have in fact all that we need to know about forcing in the definitions of x^G and $M[G]$ and these two lemmas, the truth lemma and the definability lemma. In this section we show how they are used to prove the ZFC axioms.

The empty set will be in M and so in $M[G]$, and also ω will be in M and so in $M[G]$, by 8.8.9. So since $M[G]$ is by definition a transitive set, we immediately have the axiom of extensionality, the null-set axiom, the axiom of foundation, and the axiom of infinity holding in $M[G]$ (just as in 8.4.1 for L). The other axioms require varying amounts of work.

8.9.1 Unordered pairs

We show that the pair-set axiom will always hold in $M[G]$. Any two elements of $M[G]$ will be of the form y^G and z^G for two terms $\overset{\circ}{y}, \overset{\circ}{z}$ of the language $\mathcal{L}_M$.

Let $\overset{\circ}{x}$ be the term given by
$$x = P \times \{y\} \cup P \times \{z\}$$
and we show that $\overset{\circ}{x}$ must have $x^G = \{y^G, z^G\}$. (This being absolute, we don't need to comment on *where* it will hold; it will in fact hold in any transitive set or class containing the three elements x^G, y^G and z^G.)

First it is clear that we shall have $y^G \in x^G$ since for any $p \in G$ we have $\langle p, y \rangle \in x$, so this p bears witness to $y^G \in x^G$ (we shall say that this p *forces* $\overset{\circ}{y} \in \overset{\circ}{x}$). Similarly $\overset{\circ}{z} \in \overset{\circ}{x}$ is forced by any p, and so by any $p \in G$. We must further show that these are the only members of x^G.

We know that every element of $M[G]$ is of the form t^G for some $\overset{\circ}{t}$ where $t \in M$, and we have $t^G \in x^G$ if and only if $t^G = t'^G$, for some $p \in G$ for which the pair $\langle p, t' \rangle \in x$. But we constructed x so that if this holds then t' is either y or z, so we are done.

We are using the fact that terms $\overset{\circ}{y}$ and $\overset{\circ}{t}$ are the same term (i.e. $\overset{\circ}{y} = \overset{\circ}{t}$) if and only if $y = t$, and this is almost the only thing we need to know about the terms used (except that we shall also want to know that they are absolute). However it should be clear that $y^G = t^G$ can hold for many different terms; and many other terms could have been used to give the pair-set $\{y^G, z^G\}$. For example, if the condition p is not in G, then it could be omitted from the formation of x without making any difference. This highlights an important point: we may not know, of two terms, $\overset{\circ}{t}$ and $\overset{\circ}{y}$ say, whether $t^G = y^G$ until we know all about G. So M cannot hope to know this (it could not possibly be defined in M). But M may well be in a position to know that *if* some condition p is in G, then $t^G = y^G$, and this will happen just when p forces $\overset{\circ}{t} = \overset{\circ}{y}$. That is part of the content of the definability and truth lemmas.

8.9.2 The sum-set axiom

Let us now see that the sum-set axiom will always hold in $M[G]$. Given a term $\overset{\circ}{x}$, we want a term $\overset{\circ}{y}$ such that, for any generic G,
$$y^G = \bigcup x^G = \{z^G \mid \exists u^G \in x^G (z^G \in u^G)\}$$
so that this will all hold in $M[G]$ (since it is absolute). Now $u^G \in x^G$ will hold if $\langle p, u \rangle \in x$ for some $p \in G$, and $z^G \in u^G$ if $\langle q, z \rangle \in u$ for some $q \in G$. So we shall put pairs $\langle r, z \rangle$ into y, for all conditions r which extend both p and q, whenever both the above happen; this will ensure that we get all that we want and no more. Formally,
$$y = \{\langle r, z \rangle \mid \exists p, q, u(\langle p, u \rangle \in x \wedge \langle q, z \rangle \in u \wedge r \preccurlyeq p \wedge r \preccurlyeq q)\}.$$

The quantifiers can all be bounded, and this is an absolute term and will be in M (i.e. a set in M) whenever x is.

CONSTRUCTIBLE SETS AND FORCING   161

Now to check that we get what we want: suppose $z^G \in y^G$. Then we may suppose $\langle r, z \rangle \in y$ for some $r \in G$, so we must have $\langle p, u \rangle \in x$ and $\langle q, z \rangle \in u$ with both $r \not\leq p$ and $r \not\leq q$, hence with $p, q \in G$ since G is closed upward. So we have both $u^G \in x^G$ and $z^G \in u^G$. So y^G has no more than the members we want.

Then suppose that $t^G \in \bigcup x^G$ holds; we must have $t^G \in u^G \in x^G$ for some u^G. So we may suppose $\langle p, u \rangle \in x$ and $\langle q, t \rangle \in u$ for some $p, q \in G$. But G is consistent, so some r with both $r \not\leq p$ and $r \not\leq q$ is in G, and we put $\langle r, t \rangle$ into y for all such r (whether or not they were in G, since M will not know that). But since one such (at least) will be in G, we shall have $t^G \in y^G$ and y^G will have all the members we want, and the sum-set axiom will hold in $M[G]$.

8.9.3 The power-set axiom

Next, we show that the power-set axiom holds in $M[G]$. Given $x^G \in M[G]$ we must find a term to denote its power-set; and we know that this power-set will not be absolute, so we can expect to have to work harder. But in fact it is not all that much harder. First form the set

$$Z_x = \{\langle q, t \rangle \mid \text{ for some } p \text{ with } q \not\leq p, \langle p, t \rangle \in x\}.$$

Then let $Z'_x = \{u \mid u \subseteq Z_x\}^M$ and $y = P \times Z'_x$. We shall show that y^G will be the power-set of x^G in $M[G]$.

Note that in forming Z'_x we relativized to M, that is we took only those subsets of Z_x which were in M. This is essential to ensure that Z'_x and hence y are in M, so that y^G will indeed be in $M[G]$. This is where we can claim to be working harder, this term will not be absolute for M but would change if we changed M.

However we can show what we want. First suppose that $t^G \in y^G$ and we show that $t^G \subseteq x^G$. We may suppose $\langle p, t \rangle \in y$ for some $p \in G$, and so $t \subseteq Z_x$. So any member of t^G is of the form w^G with $\langle q, w \rangle \in t$ and $q \in G$, so with $\langle q, w \rangle \in Z_x$, and hence $\langle r, w \rangle \in x$ for some r with $q \not\leq r$. But since G is closed upwards, $r \in G$ and hence $w^G \in x^G$. So we have shown $t^G \subseteq x^G$, and so $y^G \subseteq \mathcal{P}(x^G)$ (in $M[G]$, but this way round is absolute anyway).

For the other way round, we must take all subsets of x^G which are in $M[G]$ and show that they are members of y^G. For this we use the fact that every member of $M[G]$ is of the form u^G for some term $\mathring{u}$ with $u \in M$. So suppose $u^G \subseteq x^G$, and form the set

$$u' = \{\langle p, w \rangle \in Z_x \mid p \Vdash \mathring{w} \in \mathring{u}\}.$$

Because of the definability lemma, this will be a set in M, and so $\langle p, u' \rangle \in y$ for all $p \in P$ by the construction of y. Hence we must have $u'^G \in y^G$.

Now we use the fact that $u^G \subseteq x^G$ to show that $u'^G = u^G$. Suppose that $w^G \in u^G$; then also $w^G \in x^G$ and so we can assume that $\langle r, w \rangle \in x$ for some

$r \in G$ and (by the truth lemma) that $q \Vdash \overset{\circ}{w} \in \overset{\circ}{u}$ for some $q \in G$. G is consistent, so let $p \in G$ extend both q and r and we shall have $\langle p, w \rangle \in Z_x$ and $p \Vdash \overset{\circ}{w} \in \overset{\circ}{u}$ (see the note below). So $\langle p, w \rangle \in u'$ and since $p \in G$ we must have $w^G \in u'^G$. So $u^G \subseteq u'^G$.

For the other way, if $w^G \in u'^G$ then we can suppose $\langle p, w \rangle \in u'$ for some $p \in G$, and so $p \Vdash \overset{\circ}{w} \in \overset{\circ}{u}$ by the definition of u', hence by the truth lemma, $w^G \in u^G$. So $u^G = u'^G$, and we have shown that every subset of x^G which is in $M[G]$ is in y^G as required. So the power-set axiom will hold in $M[G]$.

Note that in the above proof we used the fact that if $q \Vdash \varphi$ for any formula of $\mathcal{L}_M$ then also $p \Vdash \varphi$ for any extension $p \preccurlyeq q$. This follows from the truth lemma (we leave that as an exercise); but it will also be built in when we come to give the definition which proves the definability lemma. Also when we had that both $w^G \in u^G$ and $w^G \in x^G$, we had a choice. We could *either* assume that $\langle p, w \rangle \in u$ for some $p \in G$, or that $\langle p, w \rangle \in x$ for some $p \in G$; but we cannot assume both. This is because there will be many other terms $\overset{\circ}{w}'$ for which $w^G = w'^G$ (in fact a proper class of such terms, in the sense of $M[G]$; it won't be definable in M). For each of them we shall have $q \Vdash \overset{\circ}{w} = \overset{\circ}{w}'$ for some $q \in G$; but only for some of them (a set of M, in each case) will we have $\langle r, w' \rangle \in x$, and for another set, $\langle s, w' \rangle \in u$, (for any $r, s \in P$). And there is no reason why there should be any overlap at all; in general there will not be. So we must use the truth lemma as we did above; there we chose to use a term $\overset{\circ}{w}$ for which $\langle r, w \rangle \in x$, and then since $w^G \in u^G$, this must be forced.

8.9.4 Exercises

(1) Show from the truth lemma that forcing must have the following properties, if P is non-trivial:

  (i) If $p \Vdash \varphi$ and $q \preccurlyeq p$ then $q \Vdash \varphi$.
  (ii) Suppose that whenever $q \preccurlyeq p$ and $q \neq p$, $q \Vdash \varphi$. Then $p \Vdash \varphi$.
  (iii) Suppose that there is no $q \preccurlyeq p$ such that $q \Vdash \varphi$. Then $p \Vdash \neg \varphi$.
  (iv) For any $p \in P$ and any formula φ there is some $q \preccurlyeq p$ which *decides* φ, i.e. either $q \Vdash \varphi$ or $q \Vdash \neg \varphi$; and if p does not decide φ then there are two extensions of p, one which forces φ and the other $\neg \varphi$.

(2) Show that $p \Vdash \varphi$ for any logically valid φ. Show that if $p \Vdash \psi$ for all $\psi \in \Psi$, and $\Psi \vdash \varphi$, then $p \Vdash \varphi$.

8.9.5 The replacement axiom

We need to show that if we have a definable function over $M[G]$, then the image of any member of $M[G]$ will again be a member of $M[G]$. First we note that any formula used to define a function over $M[G]$ will be a formula of the

CONSTRUCTIBLE SETS AND FORCING   163

usual language of set theory with free variables, two of which will be singled out for the functional property and the others (parameters) are assigned to elements of $M[G]$. Every element of $M[G]$ is of the form t^G for some $t \in M$, and so we can take the formula as a formula ψ of the forcing language $\mathcal{L}_M$, with the property that

$$M[G] \models \psi(y^G, z^G) \wedge \psi(y^G, w^G) \Rightarrow z^G = w^G$$

(where we have not written the parameters v_i or the elements y_i^G assigned to them, nor the universal quantifiers which are implied).

The truth lemma then tells us that we shall have

$$p \Vdash \psi(\mathring{y}, \mathring{z}) \wedge \psi(\mathring{y}, \mathring{w}) \Rightarrow \mathring{z} = \mathring{w} \text{ for some } p \in G \qquad (*)$$

(and here all the parameters y_i^G are replaced by terms $\mathring{y}_i$; we shall not mention them in the rest of this proof, but the same parameters are implied whenever ψ occurs from now on).

Now suppose that a^G is the set whose image we require. Set

$$Z_a = \{\langle q, y \rangle \mid \exists r \in P(\langle r, y \rangle \in a \wedge q \leqslant r)\}.$$

For each $\langle q, y \rangle \in Z_a$, consider all terms $\mathring{z}$ such that $q \Vdash \psi(\mathring{y}, \mathring{z})$. If there are any, there will in fact be a proper class in the sense of M, and we must cut down to a set. We use replacement in M to do this, in the shape of the rank function of M. Let $\rho(q, y)$ be the least rank of a set z for which $q \Vdash \psi(\mathring{y}, \mathring{z})$ if there are any (0 otherwise), and let

$$b = \{\langle q, z \rangle \mid \exists \langle q, y \rangle \in Z_a(q \Vdash \psi(\mathring{y}, \mathring{z}) \wedge \text{rank}(z) \leq \rho(q, y))\}^M.$$

This will be a set in M since its rank will be bounded in M (noting that Z_a is a set in M). We claim that b^G will be the image of a^G required.

First suppose that $t^G \in b^G$, then we may assume $\langle q, t \rangle \in b$ for some $q \in G$. Then some $\langle q, y \rangle \in Z_a$ with $q \Vdash \psi(\mathring{y}, \mathring{t})$. So since $q \in G$ we have $y^G \in a^G$ and $M[G] \models \psi(y^G, t^G)$, so t^G is in the image as required.

Then suppose that w^G is in the image of a^G under the function defined by ψ in $M[G]$; say $M[G] \models \psi(y^G, w^G)$ and $\langle r, y \rangle \in a$ for $r \in G$. By the truth lemma, some $q \in G$ will force $\psi(\mathring{y}, \mathring{w})$. So by the definition of b there will be some $\mathring{z}$ with $\langle q, z \rangle \in b$ and $q \Vdash \psi(\mathring{y}, \mathring{z})$.

Now we use $(*)$ and the fact that G is consistent to find a condition $s \in G$ extending p, q and r which forces $\psi(\mathring{y}, \mathring{w})$ and $\psi(\mathring{y}, \mathring{z})$ and $\psi(\mathring{y}, \mathring{z}) \wedge \psi(\mathring{y}, \mathring{w}) \Rightarrow \mathring{z} = \mathring{w}$. So using 8.9.4(2) we will have $w^G = z^G$ and $z^G \in b^G$, which completes the result.

8.9.6 The axiom of choice

We shall prove that if x^G is any set in $M[G]$, then there is a function in $M[G]$ mapping an ordinal into a set containing x^G as a subset. It will follow that x^G can be well-ordered in $M[G]$ and hence the axiom of choice holds. (This turns out to be simpler to show than any other form; given the remaining axioms, it is equivalent to any other form of the axiom of choice, see 7.2.5(1).)

Given $x \in M$ let $\{y_\beta \mid \beta < \alpha\}$ enumerate the range of x. (Remember, the $x \in M$ that we are really interested in will be relations with domain a subset of P.) By the axiom of choice in M, we can assume that this enumeration is a set in M. Now we transform this enumeration into a set in M which gives a term for the function we want in $M[G]$. First for any $y, z \in M$ let $\mathrm{pair}(y, z)$ be the set built in 8.9.1 to give the unordered pair $\{y^G, z^G\}$, namely $P \times \{y\} \cup P \times \{z\}$. Then let $\mathrm{opair}(y, z)$ be $\mathrm{pair}(\mathrm{pair}(y, y), \mathrm{pair}(y, z))$, so that if $v = \mathrm{opair}(y, z)$ then $v^G = \langle y^G, z^G \rangle$. Finally let

$$f = \{\langle p, \mathrm{opair}(\check{\beta}, y_\beta)\rangle \mid p \in P \wedge \beta < \alpha\}$$

so that f^G will contain all pairs $\langle \check{\beta}^G, y_\beta^G \rangle$ for all $\beta < \alpha$. Since $\check{\beta}$ was defined so that $\check{\beta}^G = \beta$, this means that f^G will enumerate all the possible elements of $M[G]$ which could be members of x^G, and hence $x^G \subseteq \mathrm{range}(f^G)$. As noted above, this means that the axiom of choice will hold in $M[G]$.

8.10 Forcing in practice: some models

We present in this section some specific generic extensions, and show something of how the properties of the partial ordering P can determine the properties of the generic extension $M[G]$.

8.10.1 A model with $V \neq L$

This is one of the first models given by Cohen, and is usually referred to as "adding one Cohen-generic real". We shall try to explain why. Of course, we shall have $V \neq L$ in any generic extension for which $M \neq M[G]$, since (from the work in 8.6) we know that there is only one transitive model of $V = L$ with any fixed ordinal height; if that height is α, the model is L_α, and L_α will be included in every transitive model of ZFC of height α. So even if M satisfies $V = L$, $M[G]$ cannot. But this is the simplest generic extension, and so we present it.

The partial ordering P_0 we shall use here is the set of all finite partial functions from ω into $2 = \{0, 1\}$, i.e. all functions whose domain is a finite subset of ω and which take just the values 0 and 1. If p and q are such functions, the notion of extension is exactly what you would expect: $q \preccurlyeq p$ if

CONSTRUCTIBLE SETS AND FORCING

$p \subseteq q$ (so q extends p as a function). Given this, it is clear that p and q will be incompatible only if they are both defined for some $n \in \omega$ and take different values for that n. But any condition (i.e. any such p) can be extended to two incompatible conditions, by choosing some n for which p is not defined, and extending p to $p \cup \{\langle n, 0 \rangle\}$ and $p \cup \{\langle n, 1 \rangle\}$ which are incompatible. So this is a non-trivial forcing, and we shall have $M \neq M[G]$.

Why is this forcing described as "adding a Cohen-generic real" (or sometimes just a "Cohen real")? Since the real numbers are in one-to-one correspondence with the subsets of ω, we have a new real if and only if we have a new subset of ω, and it is common to find set-theorists talking of the subsets of ω as "the reals" (and so identifying $\mathbb{R}$ and $\mathcal{P}(\omega)$). And for this forcing P_0, we can show that a set G which is P_0-generic over M is uniquely determined by a subset of ω: let $f_G = \bigcup G$ and $x_G = \{n \in \omega \mid f_G(n) = 1\}$ (so that f_G is a function and is the characteristic function of x_G). Then we can show that f_G must be total, that is dom(f_G) $= \omega$, and that G is the set of all finite partial functions from ω to 2 which are subsets of f_G, or equivalently for which $p(n) = 1$ if and only if $n \in x_G$. (See the exercise.)

8.10.2   Exercise

(1) Show that f_G must be total.
   [For each $n \in \omega$ let $D^{(n)}$ be the set $\{p \in P_0 \mid n \in \text{dom}(p)\}$. Show $D^{(n)}$ is in M and is dense, since for every condition p, if p is not already defined at n then we can extend it so that it is (and p still remains finite), so we can extend p to a condition in $D^{(n)}$. So $G \cap D^{(n)} \neq \emptyset$. Since G is consistent, f_G must be a total function, and the characteristic function of the real x_G.]
   Show that all finite subfunctions of f_G must be in G.
   [Let D^* be the set $\{p \in P_0 \mid \text{dom}(p) \in \omega\}$. Remember that $n \in \omega$ means $n = \{0, 1, \ldots, n-1\}$ and show that D^* is in M and is pre-dense; and use that G is closed upward, hence under subsets in this ordering.]

8.10.3   A model in which the continuum hypothesis is false

For this we shall add many Cohen reals all together, and so make the continuum larger; but at the same time we must be sure that we do not change $\aleph_1$ in the resulting model $M[G]$. This is an important question in most forcing applications: which cardinals of M are *preserved* (i.e. are still cardinals in $M[G]$)? We shall introduce the simplest of methods for this; but it can be hard to show which cardinals are preserved.

The forcing partial order P_κ we use here is the set of all finite partial functions from $\kappa \times \omega$ into 2, again ordered by extension in the usual sense. Here κ is an initial ordinal of M; note that the formula Init(x) is not absolute, and that is why we shall have to check more in this model. To start with we allow any initial ordinal of M; but we shall really be interested in those which are $\geq \aleph_2$ and of cofinality $> \omega$ in the sense of M.

First let us see that this forcing will add a sequence of κ new and different reals to M. For each $\alpha < \kappa$ let $f_\alpha = \{\langle n, i \rangle \mid \exists p \in G(p(\alpha, n) = i)\}$ (where G is P_κ-generic over M). Just as for P_0, conditions are incompatible only if they take opposite values for at least one pair $\langle \alpha, n \rangle \in \kappa \times \omega$. Hence the set $D^{(\alpha,n)}$ of all conditions which are defined at (α, n) will be dense. It will be in M, and so G meets this set and hence f_α will be a total function on ω and will be the characteristic function of a real $x_\alpha \subseteq \omega$.

Why is this x_α new to M? Well, if $y \in M$ is a subset of ω, we can show that $y \neq x_\alpha$ by an application of the truth lemma. We can suppose we have translated the description of x_α given above, into a term of the forcing language $\mathring{t}_\alpha$ which denotes x_α in any $M[G]$. Then if $x_\alpha = y$ in some $M[G]$, the truth lemma tells us that it must be forced: some condition $p \in P_\kappa$ must force the formula $\mathring{t}_\alpha = \check{y}$. But this p is finite, so we can find $n \in \omega$ for which $p(\alpha, n)$ is not defined. Now extend p to $q \in P$ with $q(\alpha, n) = 0$ if $n \in y$, and $q(\alpha, n) = 1$ if $n \notin y$. This q will ensure that $x_\alpha \neq y$, since if $q \in G$ then x_α and y will differ on n. So since $q \in G \Rightarrow p \in G$ we cannot have $p \Vdash \mathring{t}_\alpha = \check{y}$, and in fact we have shown that every condition will force $\mathring{t}_\alpha \neq \check{y}$.

A very similar argument will show that every condition will force $\mathring{t}_\alpha \neq \mathring{t}_{\alpha'}$ for any $\alpha \neq \alpha' < \kappa$. We leave it as an exercise to show that the function correlating each $\alpha < \kappa$ with x_α will be in $M[G]$; given that, it is clear that if $\kappa > \aleph_1$ in M then the continuum hypothesis will fail in $M[G]$, unless some initial ordinal in M is no longer an initial ordinal in $M[G]$. We show that for this P_κ, that does not happen, using the important notion of the *countable chain condition*:

Definition 8.10.4 A poset P satisfies the *countable chain condition* (the c.c.c.), if every antichain in P is countable.

(The slightly contradictory use of the word *chain* here has its origin in earlier uses of this condition, when the partial orderings were not so much in consideration, and what is now an antichain in the partial order was called a chain in another situation; some writers use *countable antichain condition*, (c.a.c.) for the same thing.)

Let us see that P_κ satisfies this condition. We have already noted that for $p, q \in P_\kappa$, $p \perp q$ if and only if $p(\alpha, n)$ and $q(\alpha, n)$ are both defined and different, for some $\alpha < \kappa$ and $n < \omega$. It is this fact, together with the fact that the domains of p and q are finite, which restricts the possibilities for large sets of incompatible conditions. Since it is a useful combinatorial fact in its own right, we set out the lemma needed: it is usually called the *delta-system lemma*.

CONSTRUCTIBLE SETS AND FORCING

Lemma 8.10.5 *Let $\mathcal{A}$ be an uncountable collection of finite sets. Then $\mathcal{A}$ has an uncountable subset $\mathcal{B}$ which forms a delta-system, that is, there is a finite set d such that for any two members $b \neq b'$ from $\mathcal{B}$ we have $b \cap b' = d$.*

The set d is sometimes called the *root* of the delta-system, and the picture one may form is that each member of the delta-system contains the root, but then branches out in a different direction from all other members. The root may be empty, in which case we would have a disjoint system; but that will not always be possible. The delta-system lemma may be said to show how close to a disjoint system one can get.

Proof First we may restrict $\mathcal{A}$ to an uncountable subset $\mathcal{A}_1$ such that each member of $\mathcal{A}_1$ has the same size (n_1, say). This is because there are only countably many possible sizes, as all members of $\mathcal{A}$ are finite, so that if we put $\mathcal{A}^{(n)} = \{a \in \mathcal{A} \mid |a| = n\}$ and each such $\mathcal{A}^{(n)}$ were countable, then $\mathcal{A}$ would be a countable union of countable sets and so countable.

Next let c be any finite set, and consider $\{a \in \mathcal{A}_1 \mid c \subseteq a\}$. For some c this set will be uncountable (e.g. for $c = \emptyset$); let $\mathcal{C}$ be the set of all such c. Formally,

$$\mathcal{C} = \{c \mid |\{a \in \mathcal{A}_1 \mid c \subseteq a\}| > \aleph_0\}.$$

Each $c \in \mathcal{C}$ is finite and of limited size (less than n_1). So $\mathcal{C}$ must have a maximal member; let d be such a maximal member, and let $\mathcal{A}_2 = \{a \in \mathcal{A}_1 \mid d \subseteq a\}$ (so by the definition of $\mathcal{C}$, $\mathcal{A}_2$ is uncountable). (We could have $d = \emptyset$, but in general we won't.)

This d will have the property that for any element $e \notin d$, $d \cup \{e\} \subseteq a$ can hold for at most countably many $a \in \mathcal{A}_2$; otherwise d would not be maximal in $\mathcal{C}$. Hence for any set E disjoint from d, the set of $a \in \mathcal{A}_2$ for which $a \cap E \neq \emptyset$ can have cardinal at most $|E| + \aleph_0$. We can now use this fact to build our delta-system $\mathcal{B}$. We will have d as root, and we assume that $\mathcal{A}_2$ is well-ordered so that we can always pick the next element of $\mathcal{A}_2$ with the required property, if there are any at all.

So take a_0 as the first element of $\mathcal{A}_2$, and assume for induction that we have found a_β for $\beta < \gamma$ so that $\{a_\beta \mid \beta < \gamma\}$ forms a delta-system with root d. Then provided $\gamma < |\mathcal{A}_2|$ we can continue: put $E = \bigcup_{\beta < \gamma}(a_\beta - d)$. The next element a_γ must be disjoint from E. There must be such an element (and we can take the first such as a_γ, and so continue the inductive definition), since $|E| \leq |\gamma| + \aleph_0$, which is less than $|\mathcal{A}_2|$, since $\mathcal{A}_2$ is uncountable. So there are less than $|\mathcal{A}_2|$ members of $\mathcal{A}_2$ which meet E, and there must be some which do not, which is all we require to continue the induction. This induction can continue at least for uncountably many steps, and we take $\mathcal{B}$ as the set of elements given by this induction. $\square$

Note the use of the axiom of choice in this proof. The proof can be refined to give further information about the size of the delta-system, and to apply to other cardinals than just finite and uncountable. The name "delta-system lemma" will often refer to these extensions; some are in the exercises. But the simplest version, which we have given above, is sufficient for the use we make of it:

Lemma 8.10.6 *For any κ, the partial order P_κ satisfies the countable chain condition.*

Proof Suppose that $\mathcal{A} \subseteq P_\kappa$ is an antichain. We must show that $\mathcal{A}$ is countable, and we use the delta-system lemma to derive a contradiction from the assumption that it is not. We apply the delta-system lemma to the set $\mathcal{A}' = \{\operatorname{dom}(p) \mid p \in \mathcal{A}\}$. Each of these domains is finite, so if $\mathcal{A}'$ is uncountable, we get an uncountable set $\mathcal{B}' \subseteq \mathcal{A}'$ which is a delta-system; let it have root d, and let $\mathcal{B} = \{p \in \mathcal{A} \mid \operatorname{dom}(p) \in \mathcal{B}'\}$. This $\mathcal{B}$ must also be uncountable, but for any two members $p, q \in \mathcal{B}$ we have $\operatorname{dom}(p) \cap \operatorname{dom}(q) = d$ and d is finite. But if p and q are incompatible, they must disagree in value for some element of d. If d has n members there are only 2^n ways to assign the values 0 and 1 to each member of d. So $\mathcal{B}$ can have at most 2^n members, otherwise at least two must agree on d and so would be compatible, which contradicts the assumption that $\mathcal{A}$, and hence $\mathcal{B}$, is an antichain.

This contradiction means that $\mathcal{A}$ must have been countable, and so P_κ has the c.c.c as claimed. □

Now we show how this is used to prove that cardinals (i.e. initial ordinals) and also cofinalities are unchanged between M and $M[G]$, when G is P_κ-generic over M, for any κ:

Lemma 8.10.7 *If P has the c.c.c. and $f \in M[G]$ is a function with domain an infinite ordinal α and co-domain an ordinal β, then there is a set $Y \in M$ such that $\operatorname{range}(f) \subseteq Y$ and in M, $|Y| = |\alpha|$.*

Proof There must be a term $\overset{\circ}{f}$ such that f is f^G, and since ordinals are the same in M and in $M[G]$, and $f^G : \alpha \to \beta$ holds in $M[G]$, the truth lemma tells us that some $p \in P$ must force the corresponding formula $\overset{\circ}{f} : \check{\alpha} \to \check{\beta}$. Now we can work in M: we have $p \Vdash \forall \delta \in \check{\alpha}(\overset{\circ}{f}(\delta) < \check{\beta})$, so for each $\delta < \alpha$ and each generic G with $p \in G$ we must have a $\gamma < \beta$ such that $f^G(\delta) = \gamma$ and hence a condition $q \preccurlyeq p$ such that $q \Vdash \overset{\circ}{f}(\check{\delta}) = \check{\gamma}$.

So for each $\delta < \alpha$ we can (using the definability lemma) form the set Y_δ in M:

$$Y_\delta = \{\gamma < \beta \mid \exists q \preccurlyeq p(q \Vdash \overset{\circ}{f}(\check{\delta}) = \check{\gamma})\},$$

and this set must contain the actual value which f^G takes for argument δ (even though M may have no way to tell which of these possible values will in fact be taken, until p is extended to decide which).

CONSTRUCTIBLE SETS AND FORCING   169

This set Y_δ must be countable in M, because if $q \Vdash \overset{\circ}{f}(\check\delta) = \check\gamma$ and $q' \Vdash \overset{\circ}{f}(\check\delta) = \check\gamma'$ with $\gamma \neq \gamma'$ and both $q, q' \preccurlyeq p$, then q and q' must be incompatible (since p forces $\overset{\circ}{f}$ to denote a function). So choosing one condition $q \preccurlyeq p$ for each $\gamma \in Y_\delta$ will give an antichain in P, and P has the c.c.c.

Now the set $Y = \bigcup_{\delta < \alpha} Y_\delta$ will be in M and will have $|Y| = |\alpha|$ since α is infinite; and from its construction we have range$(f^G) \subseteq Y$ for any G, as required. □

Corollary 8.10.8 *All cardinalities and cofinalities are preserved between M and $M[G]$, if P has the c.c.c.*

Proof All cardinalities and cofinalities are decided by either one-to-one or onto functions $f : \alpha \to \beta$ for infinite ordinals α and β, and the lemma shows that if we have such a function in $M[G]$ then (since $\alpha \times \alpha \sim \alpha$) we must already have a function in M with the same properties. For example, if f is a bijection, then we must have $\beta \subseteq Y$ for the set $Y \in M$ given by the lemma; but this shows that $\alpha \sim \beta$ in M. If f is cofinal in β, then Y must be cofinal in β and so there must be some cofinal function $g : \alpha \to \beta$ in M. □

8.10.9 A model in which cardinals are collapsed

Just to show that the work done above, to confirm that cardinals were not changed, was essential, we give a model in which they definitely are changed. Let $P_{\omega,\beta}$ be the partial ordering given by all finite partial functions from ω into β, where β is any ordinal of M, with $P_{\omega,\beta}$ ordered by the usual notion of extension. Here we are thinking of β as an uncountable initial ordinal of M.

Suppose that G is P-generic over M, and set $f_G = \bigcup G$. Just as for P_0 in 8.10.1, we shall have that f_G is a function with domain ω. What will be its range? Well, if we let $D_{(\alpha)} = \{p \in P \mid \alpha \in \text{range}(p)\}$, it is easy to see that $D_{(\alpha)}$ will be dense for every $\alpha < \beta$, since we can always extend any condition to one in $D_{(\alpha)}$ by adding $\langle n, \alpha \rangle$ to p for some n not already in dom(p). So every generic G must meet every $D_{(\alpha)}$, and hence the range of f_G will be the whole of β—whatever β we chose in M.

So if β is chosen as any uncountable initial ordinal in M, in $M[G]$ there will be a function mapping ω onto β, so that in $M[G]$, β is countable.

Note that this result is not very surprising in view of the fact that we started with a ground model M which is countable, so that every ordinal β of M is in fact countable, although M will not always have an enumeration of β. But it is clear that it required some delicacy to add such an enumeration whilst still getting a model of ZFC. And this case does show that we cannot expect to find generic subsets in all cases, if we do not start with a countable ground model.

8.10.10   Exercises

(1) In the notation of 8.10.3, show that the function mapping $\alpha < \kappa$ to x_α will be in $M[G]$.
   [Write a term to give $\{\langle \alpha, x_\alpha \rangle \mid \alpha < \kappa\}$.]

(2) Computing $2^{\aleph_0}$ in $M[G]$. We did not show what this value was in 8.10.3; we only showed that it was at least κ and (in 8.10.8) that κ is still an initial ordinal. We can compute an upper bound as follows:
   Show that if $u^G \subseteq \omega$ in $M[G]$ then we can assume that every member of u is of the form $\langle p, \check{n} \rangle$ with $p \in P$ and $n \in \omega$.
   Hence show that there is a set $y \in M$ of size 2^κ in M such that whenever $u^G \subseteq \omega$ in $M[G]$ then some condition forces $\mathring{u} \in \mathring{y}$.
   [Note that in M, $|P_\kappa| = \kappa$; and use the proof of the power-set axiom as in 8.9.3.]
   Deduce that if $\lambda = 2^\kappa$ in M then $2^{\aleph_0} \leq \lambda$ in $M[G]$.
   For a better upper bound, use the following: say that $\mathring{u}$ is a *sparse* name for a subset of ω if $u = \bigcup \{A_n \times \{\check{n}\} \mid n \in \omega\}$ where for each n, A_n is an antichain in P_κ. Show that there are κ^ω many sparse names in M (the cardinal κ^ω being computed in M).
   Then show that for any name $\mathring{t}$ there is a sparse name $\mathring{t}'$ with the property that whenever $t^G \subseteq \omega$ then $t^G = t'^G$; and deduce that if $\kappa^\omega = \kappa$ in M then in $M[G]$, we have $2^\omega = \kappa$.
   [Take $t' = \bigcup \{A_n \times \{\check{n}\} \mid n \in \omega\}$ where A_n is an antichain in P with the property that $\forall p \in A_n (p \Vdash \check{n} \in \mathring{t})$, and is maximal with this property. Note that this can be defined in M.]

(3) Say that P has the λ-*chain condition* if every antichain in P has cardinal $< \lambda$. (So the c.c.c. is the $\aleph_1$-chain condition.)
   Show that if $P \in M$ has the λ-chain condition in M, and λ is regular in M, then every cardinal of $M \geq \lambda$ will be preserved.
   [Prove the equivalent of lemma 8.10.7.]

(4) Show directly that the forcing partial order $P_{\omega,\beta}$ does not satisfy the c.c.c.
   [Find an antichain of size β in M by considering the conditions $\{\{\langle n, \alpha \rangle\} \mid \alpha < \beta\}$.]

(5) Stronger versions of the delta-system lemma. Let λ be a regular cardinal and $\alpha < \lambda$. Show that if we start with a collection $\mathcal{A}$ of λ many sets, each of size $< \alpha$, and we have $\delta^{<\alpha} < \lambda$ for all $\delta < \lambda$, then there must be a subset $\mathcal{B} \subseteq \mathcal{A}$ with $|\mathcal{B}| = \lambda$ which forms a delta-system.
   [Here the delta-system will have root of cardinal $< \alpha$, and the hypothesis on cardinal powers, $\delta^{<\alpha} < \lambda$ for all $\delta < \lambda$, says that any set of size less than λ will have less than λ subsets of size $< \alpha$. Show that these facts are just what is needed to carry through the proof given in 8.10.5 replacing "uncountable" by "size λ" and "finite" by "size $< \alpha$".]

CONSTRUCTIBLE SETS AND FORCING

The hypothesis that λ is regular is needed for the first step, where we restrict $\mathcal{A}$ to some subset all of the same size.]

(6) Use the stronger version of the delta-system lemma in (5) to show that if β is regular in M, any antichain in $P_{\omega,\beta}$ which is in M must have cardinal $\leq \beta$ in M. Hence show that cardinals of M which are $> \beta$ in M will be preserved.

(7) Show that if all cofinalities are preserved between M and $M[G]$, then all cardinalities are also preserved; and that if all regular cardinals of M are still regular in $M[G]$ then all cardinals and cofinalities are preserved.

8.11 Proofs of the definability and truth lemmas

These were stated in 8.8.7 and 8.8.8. To prove them we shall set up a definition of forcing for the language $\mathcal{L}_M$ which we shall write as $\Vdash^*$, and for which we can prove enough of the properties we want (some of the properties of 8.9.4). We shall show that this definition is absolute for M. This relation will not be exactly the forcing relation $\Vdash$ defined in 8.8.6; it will in fact be a stronger relation (it will hold less often; it is sometimes called *strong forcing*, and the relation we have been working with so far is *weak forcing*). We shall show that we can define $\Vdash$ from $\Vdash^*$ by the definition:

$$p \Vdash \varphi \text{ if and only if } p \Vdash^* \neg\neg\varphi,$$

and then we shall be able to prove the truth lemma, with $M[G]$ defined as in 8.8.6.

8.11.1 The definition for atomic formulas

Most of the work has to go into this step. The atomic formulas we are concerned with are of the form $\mathring{y} \in \mathring{x}$ and $\mathring{y} = \mathring{x}$ with x and y in M, and the definition will proceed by induction on the ranks of the sets x and y, and simultaneously for both $=$ and $\in$. We shall need the notion of a *dense set of* $p' \preccurlyeq p$, which is a set of conditions all below p such that every extension of p can be further extended to be in the set.

The basic definition is (for $\in$):

$p \Vdash^* \mathring{y} \in \mathring{x}$ if and only if,
for some z and q with $\langle q, z \rangle \in x$ and $p \preccurlyeq q$,
we have $p \Vdash^* \mathring{y} = \mathring{z}$;

and this needs the subsidiary definition that if $\alpha = \max(\mathrm{rank}(x), \mathrm{rank}(y))$, then (for =):

$p \Vdash^* \overset{\circ}{y} = \overset{\circ}{x}$ if and only if,

for all $z \in M$ with $\operatorname{rank}(z) < \alpha$ and all $p' \preccurlyeq p$,

if $p' \Vdash^* \overset{\circ}{z} \in \overset{\circ}{y}$ then for a dense set of $p'' \preccurlyeq p'$, $p'' \Vdash^* \overset{\circ}{z} \in \overset{\circ}{x}$, and

if $p' \Vdash^* \overset{\circ}{z} \in \overset{\circ}{x}$ then for a dense set of $p'' \preccurlyeq p'$, $p'' \Vdash^* \overset{\circ}{z} \in \overset{\circ}{y}$.

We have claimed that this is a definition by induction on the rank; let us first see that this is so, in other words that it is well-founded. Let $\alpha = \max(\operatorname{rank}(x), \operatorname{rank}(y))$ as above. In order to decide whether $p \Vdash^* \overset{\circ}{y} \in \overset{\circ}{x}$ we need to know all cases of $p' \Vdash^* \overset{\circ}{y} = \overset{\circ}{z}$ for z in the range of x, and hence with $\operatorname{rank}(z) < \operatorname{rank}(x)$; and to decide $p' \Vdash^* \overset{\circ}{y} = \overset{\circ}{z}$ we need to know all cases of $p'' \Vdash^* \overset{\circ}{u} \in \overset{\circ}{z}$ and $p'' \Vdash^* \overset{\circ}{u} \in \overset{\circ}{y}$ with $\operatorname{rank}(u) < \alpha$. For the cases $p'' \Vdash^* \overset{\circ}{u} \in \overset{\circ}{z}$ the maximum of the ranks must now be less than α. For the other cases, we certainly cannot return to the question whether $p'' \Vdash^* \overset{\circ}{y} \in \overset{\circ}{x}$; but we could conceivably come to the question $p'' \Vdash^* \overset{\circ}{x} \in \overset{\circ}{y}$, if y had greater rank than x. In that case we just go one step further to see that the definitions are indeed well-founded. At the next step we shall be asking to know all cases of $p''' \Vdash^* \overset{\circ}{z} = \overset{\circ}{u}$ for $z \in \operatorname{range}(y)$ and hence for $\operatorname{rank}(z) < \alpha$, so the maximum rank will now decrease, and y will disappear from the questions asked for this next step. All the questions to be decided from this point on will have maximum rank $< \alpha$, and that is how we know the definition is well-founded.

Of course it is more natural to think of the definition proceeding upward through the ranks: first, decide all questions with maximum rank 0 (there aren't many!), then all questions with maximum rank 1, etc. Every question will be decided eventually, and for each question, only a set in the sense of M of questions has to be asked.

Note that this definition immediately gives the property that extensions force more, i.e. if $p \Vdash^* \overset{\circ}{y} \in \overset{\circ}{x}$ and $q \preccurlyeq p$ then $q \Vdash^* \overset{\circ}{y} \in \overset{\circ}{x}$, and similarly for =.

8.11.2   Absoluteness of these definitions

We can now apply the methods used in 8.4 above to write two formulas, say $\varphi_\in(\alpha, x, y, p, P)$, which will express the relation

$$\operatorname{rank}(x) < \alpha \wedge \operatorname{rank}(y) < \alpha \wedge (p \Vdash^* \overset{\circ}{x} \in \overset{\circ}{y}),$$

and $\varphi_=(\alpha, x, y, p, P)$, which will express

$$\operatorname{rank}(x) < \alpha \wedge \operatorname{rank}(y) < \alpha \wedge (p \Vdash^* \overset{\circ}{x} = \overset{\circ}{y}),$$

and to show that these are absolute for M. The quantifiers over terms will be bounded to a set in M when $\alpha \in M$, and this is the key fact which ensures

CONSTRUCTIBLE SETS AND FORCING

this absoluteness.

The definitions by transfinite induction that we have given so far have been for terms rather than for formulas of this sort. To see that we can adapt all the work on transfinite induction and absoluteness to formulas in the way we want it here, it is simplest to think of defining by induction on α, for $\alpha \in M$, the sets of quadruples

$$\{\langle x, y, p, P\rangle \mid \mathrm{rank}(x) < \alpha \wedge \mathrm{rank}(y) < \alpha \wedge (p \Vdash^* \overset{\circ}{x} \in \overset{\circ}{y})\} = F_\in(\alpha), \text{ say, and}$$

$$\{\langle x, y, p, P\rangle \mid \mathrm{rank}(x) < \alpha \wedge \mathrm{rank}(y) < \alpha \wedge (p \Vdash^* \overset{\circ}{x} = \overset{\circ}{y})\} = F_=(\alpha).$$

The methods of 8.2 and 8.4 can then be applied to show that these are absolute.

8.11.3   The definition for compound formulas

We use the following inductive steps:

(i) $p \Vdash^* \neg \varphi$ if and only if, for every q extending p, not $q \Vdash^* \varphi$;
(ii) $p \Vdash^* \varphi \wedge \psi$ if and only if both $p \Vdash^* \varphi$ and $p \Vdash^* \psi$;
(iii) $p \Vdash^* \exists v \varphi(v)$ if and only if for some $x \in M$, $p \Vdash^* \varphi(\overset{\circ}{x})$.

We give the definition (ii) in this form (for conjunction) because it is simplest; the definitions for disjunctions and for implications can be worked out from this and (i), and are left as exercises. Note that the form of (i) immediately makes this forcing consistent; we can never force a contradiction (simply because we always count p as extending itself). Also these definitions preserve the property that extensions force more.

At this stage we should note what we are doing: for each closed formula φ of the forcing language (i.e. of $\mathcal{L}_M$), say with terms $\overset{\circ}{x}_1, \ldots, \overset{\circ}{x}_n$, we are defining a formula, say $\mathrm{Force}_\varphi(v_1, \ldots, v_n, v_{n+1}, v_{n+2})$ such that for any $x_1, \ldots, x_n \in M$ and $p \in P$ with $P \in M$ a partial order, we have the equivalence:

$$p \Vdash^* \varphi(\overset{\circ}{x}_1, \ldots, \overset{\circ}{x}_n)$$

if and only if:

Force_φ holds when $x_1, \ldots, x_n, p, P$ are assigned to $v_1, \ldots, v_n, v_{n+1}, v_{n+2}$.

Absoluteness for M for this formula means that the final "holds" here can be taken either in the universe or in the model M without making any difference.

8.11.4   Proof of the truth lemma

We now work in the universe, and assume that we have a set $G \subseteq P$ which is P-generic over M, and make the definitions

$$x^G = \{y^G \mid \exists p \in G(\langle p, y\rangle \in x)\}; \quad M[G] = \{x^G \mid x \in M\}.$$

(The first definition is by induction on rank, and is to be restricted to $x \in M$.)

We prove the truth lemma by induction on the formulas, and most of the work must go into the basis, the case of atomic formulas. These are the formulas $\mathring{y} \in \mathring{x}$ and $\mathring{y} = \mathring{x}$ for $x, y \in M$, and we must show that

$$y^G \in x^G \text{ if and only if, for some } p \in G,\ p \Vdash^* \mathring{y} \in \mathring{x}$$

and

$$y^G = x^G \text{ if and only if, for some } p \in G,\ p \Vdash^* \mathring{y} = \mathring{x}.$$

We show this by induction on $\alpha = \max(\operatorname{rank}(y), \operatorname{rank}(x))$. The induction hypothesis we use is that for any $u, w \in M$ both of rank $< \alpha$,

$$u^G = w^G \text{ if and only if, for some } p \in G,\ p \Vdash^* \mathring{u} = \mathring{w}.$$

First suppose that $p \in G$ and $p \Vdash^* \mathring{y} = \mathring{x}$. We show that $y^G = x^G$. We know that $y^G = \{u^G \mid \langle q, u \rangle \in y \wedge q \in G\}$; for these u we have $\operatorname{rank}(u) < \alpha$. Given such a u^G we must show $u^G \in x^G$, using the basic definition (from 8.11.1). So take $q' \in G$ extending both p and q (where $\langle q, u \rangle \in y$).

We have $q' \Vdash^* \mathring{u} \in \mathring{y}$, with $q' \preccurlyeq p$, so the basic definition (for $=$) applied to $p \Vdash^* \mathring{y} = \mathring{x}$ tells us that for a dense set of $q'' \preccurlyeq q'$, $q'' \Vdash^* \mathring{u} \in \mathring{x}$. One of this dense set must be in G since q' is, so we get $q'' \in G$ such that $q'' \Vdash^* \mathring{u} \in \mathring{x}$. So again by the basic definition, for some $\langle s, w \rangle \in x$ we have $q'' \preccurlyeq s$ and $q'' \Vdash^* \mathring{u} = \mathring{w}$. Clearly $\operatorname{rank}(w) < \alpha$ also, so by the induction hypothesis $u^G = w^G$. But $w^G \in x^G$ by the definition of x^G so we have shown $u^G \in x^G$ as required, and hence $y^G \subseteq x^G$. Similarly we shall have $x^G \subseteq y^G$ and so $x^G = y^G$.

Note that in the proof of this step, we have shown that for u with $\operatorname{rank}(u) < \alpha$, if $q'' \in G$ and $q'' \Vdash^* \mathring{u} \in \mathring{x}$ then $u^G \in x^G$; and the converse (again for $\operatorname{rank}(u) < \alpha$) is similar. This proves most of the $\in$ case.

Now we assume $y^G = x^G$, and we must show that for some $p \in G$, $p \Vdash^* \mathring{y} = \mathring{x}$.

Suppose that it is not the case that $p \Vdash^* \mathring{y} = \mathring{x}$ for some p: applying the basic definition we must then have some $z \in M$ with $\operatorname{rank}(z) < \alpha$, and some $p' \preccurlyeq p$ such that either (i) $p' \Vdash^* \mathring{z} \in \mathring{y}$, but the set of $p'' \preccurlyeq p'$ such that $p'' \Vdash^* \mathring{z} \in \mathring{x}$ is not dense below p'; or (ii) the same with y and x reversed.

In case (i), for the set to be *not* dense, p' must have an extension p'' for which every extension does not force $\mathring{z} \in \mathring{x}$, and using 8.11.3(i) this is exactly when $p'' \Vdash^* \mathring{z} \notin \mathring{x}$; and also if $p'' \in G$ then $z^G \notin x^G$. Similarly for case (ii), $p'' \Vdash^* \mathring{z} \notin \mathring{y}$.

So consider the set D of conditions $p \in P$ such that either $p \Vdash^* \mathring{x} = \mathring{y}$, or

(i') for some z with $\operatorname{rank}(z) < \alpha$, $p \Vdash^* \mathring{z} \in \mathring{y}$ and $p \Vdash^* \mathring{z} \notin \mathring{x}$; or (ii′) the same with y and x reversed.

This set D will be in M by definability, and will be dense by the argument above, which showed that if it is not the case that $p \Vdash^* \overset{\circ}{x} = \overset{\circ}{y}$ then either (i′) or (ii′) must hold for some extension. So G meets D; suppose $p \in D \cap G$. If (i′) holds for p then, since $p \in G$ we will have $z^G \in y^G$ but $z^G \notin x^G$, and hence $y^G \neq x^G$, and similarly for case (ii′). So we have shown that if $x^G = y^G$ then some $p \in G$ must satisfy $p \Vdash^* \overset{\circ}{x} = \overset{\circ}{y}$ and we have completed the induction step for $=$.

The induction step for $\in$ now needs only the observation that if $\mathrm{rank}(y) = \alpha$ then we have proved by the step above that $y^G = z^G$ if and only if for some $p \in G$, $p \Vdash^* \overset{\circ}{y} = \overset{\circ}{z}$, whenever $\mathrm{rank}(z) \leq \alpha$. And this is all that is missing from the proof of the $\in$ case, as included in the proof of the $=$ case.

Now that we have the basis for the induction, the truth lemma follows for compound formulas very easily. We shall show for all formulas φ:

$$\exists p \in G (p \Vdash^* \varphi(\overset{\circ}{x}_1, \ldots, \overset{\circ}{x}_n)) \Leftrightarrow M[G] \models \varphi(x_1^G, \ldots, x_n^G). \qquad (*)$$

Assume as induction hypothesis that $(*)$ holds for φ, and that for some $p \in G$ we have $p \Vdash \neg \varphi$ (here omitting to write the constant symbols and corresponding denotations). Then we must have $M[G] \models \neg \varphi$ since otherwise we would have $M[G] \models \varphi$ and so $r \Vdash^* \varphi$ for some $r \in G$ (by the induction hypothesis) and a contradiction from a common extension of p and r. And if $M[G] \models \neg \varphi$, no $p \in G$ can have $p \Vdash^* \varphi$. But the set of conditions r which decide φ (i.e. which either force φ or force $\neg \varphi$) is dense (immediately from the definition for forcing $\neg \varphi$). So G meets this set, and hence some $p \in G$ has $p \Vdash^* \neg \varphi$. So we have proved $(*)$ for $\neg \varphi$.

The induction step for formulas $\varphi \wedge \psi$ is trivial, since both forcing and satisfaction treat this case alike. The induction step for $\exists v \varphi(v)$ is also straightforward, using the fact that $M[G] = \{x^G \mid x \in M\}$, so that $M[G] \models \exists v \varphi(v)$ if and only if $M[G] \models \varphi(x^G)$ for some $x \in M$. So the induction hypothesis gives this equivalent to: some $p \in G$ has $p \Vdash^* \varphi(\overset{\circ}{x})$, and hence by 8.11.3(iii), equivalent to $p \Vdash^* \exists v \varphi(v)$ as required.

The statement of the truth lemma in 8.8.7 has weak forcing, $\Vdash$, in place of the strong forcing $\Vdash^*$ in $(*)$. But using 8.11.3(i) gives: $p \Vdash^* \neg\neg \varphi$ if and only if the set of $r \preccurlyeq p$ such that $r \Vdash^* \varphi$ is dense below p (since $p \Vdash^* \neg\neg \varphi$ if and only if no extension $q \preccurlyeq p$ has $q \Vdash^* \neg \varphi$, so every extension $q \preccurlyeq p$ has an extension $r \preccurlyeq q$ with $r \Vdash^* \varphi$).

Hence if $p \in G$ and $p \Vdash^* \neg\neg \varphi$ then some $r \in G$ must have $r \Vdash^* \varphi$ (since G meets all dense sets in M). So $(*)$ is equivalent to

$$\exists p \in G(p \Vdash^* \varphi(\overset{\circ}{x}_1, \ldots, \overset{\circ}{x}_n)) \Leftrightarrow M[G] \models \varphi(x_1^G, \ldots, x_n^G),$$

simply from the definition of generic.

This completes the proof of the truth lemma. More properties of forcing are indicated in the exercises, but we do not try to survey the many uses of forcing that have been made. One important extension of the work is to *iterated forcing*, and exercises (5)–(7) below only hint at the many uses of this.

8.11.5 Exercises

(1) Many presentations of forcing will assume that every forcing partial order has a largest element $\mathbb{I}$, so that every element extends this $\mathbb{I}$. (For all the P we have used, $\emptyset$ will be the largest element.) Show that if we assume this, then names for the members of M in $M[G]$ can be given by:

$$x' = \{\langle \mathbb{I}, y' \rangle \mid y \in x\}$$

and then $\check{x}$ can be taken as $\check{x} = \overset{\circ}{x}'$ (compare 8.8.9).

(2) A maximal principle for weak forcing. We have in 8.11.3(iii) that for strong forcing, $\Vdash^*$,

$$p \Vdash^* \exists v \varphi(v) \text{ if and only if for some } x \in M, \ p \Vdash^* \varphi(\overset{\circ}{x}).$$

Show that this holds also for weak forcing, $\Vdash$.

[Use the fact that $p \Vdash \exists v \varphi(v)$ if and only if the set

$$Q_p = \{r \in P \mid r \not\preccurlyeq p \wedge r \Vdash^* \exists v \varphi(v)\}$$

is dense below p. Suppose this holds and let A_p be a maximal antichain in Q_p, and for each $r \in A_p$ let $\overset{\circ}{x}_r$ be a term such that $r \Vdash^* \varphi(\overset{\circ}{x}_r)$. Then define

$$x'_r = \{\langle q, y \rangle \mid \exists q'(\langle q', y \rangle \in x_r \wedge q \preccurlyeq q' \wedge q \preccurlyeq r)\},$$

and let $x = \bigcup \{x'_r \mid r \in A_p\}$. This x will satisfy $p \Vdash \varphi(\overset{\circ}{x})$; suppose that $p \in G$ where G is generic, then some $r \in A_p \cap G$. Show that $x^G = x_r^G$. Note the use of the axiom of choice in this proof.]

(3) Embeddings of forcing partial orders. Suppose P and Q are two partial orderings (and we shall use the same symbol $\preccurlyeq$ for both partial orderings); an embedding $i : P \to Q$ is a *complete embedding* if

  (i) $\forall p, q \in P(p \preccurlyeq q \Rightarrow i(p) \preccurlyeq i(q))$, and
  (ii) if A is a maximal antichain in P then its image $i''A$ is a maximal antichain in Q.

It is a *dense embedding* if also

  (iii) $i''P$ is dense in Q.

Show that if $i : P \to Q$ is complete and H is Q-generic then $i^{-1}(H)$ is P-generic, and hence $M[i^{-1}(H)] \subseteq M[H]$.

Show also that if i is dense and G is P-generic and we define $i * G$ as the upward closure of $i''G$ in Q (i.e. $i * G = \{q \in Q \mid \exists p \in G(i(p) \preccurlyeq q\})$, then $i * G$ is Q-generic, and $i^{-1}(i * G) = G$ so that $M[G] = M[H]$.

(4) Separative partial orderings. A partial ordering P is *separative* if

$$\forall p, q \in P(p \not\preccurlyeq q \Rightarrow \exists r \preccurlyeq p(r \perp q)).$$

Show that P is separative if $p \preccurlyeq q \Leftrightarrow p \Vdash \check{q} \in \check{G}$.

Given any partial ordering P, we form the *separative quotient* of P by forming the equivalence relation on P:

$$p \approx q \Leftrightarrow \forall r \in P(r \| p \Leftrightarrow r \| q)$$

and then $[p] = \{q \mid q \approx p\}$ and $P/\approx = \{[p] \mid p \in P\}$ is the separative quotient, ordered by $[p] \preccurlyeq [q] \Leftrightarrow \forall r \in P(r \preccurlyeq p \Rightarrow r \| q)$.

Show that $P/\approx$ is a separative partial order, and that the embedding $i: P \to P/\approx$ given by $i(p) = [p]$ is a dense embedding. Hence all generic extensions can be given by separative partial orders.

(5) Product forcing. For two partial orders P and Q we define $P \times Q$ to have the ordering

$$\langle p, q \rangle \preccurlyeq \langle p', q' \rangle \Leftrightarrow p \preccurlyeq_P p' \wedge q \preccurlyeq_Q q'.$$

Show that G is $P \times Q$-generic if and only if $G = G_1 \times G_2$ where G_1 is P-generic over M and G_2 is Q-generic over $M[G_1]$.

[This is the simplest form of *iteration* of forcings.]

(6) General products. Suppose that P_α is a partial ordering for each $\alpha < \delta$, each with a largest element $\mathbb{I}_{P_\alpha}$. For an element $p \in \bigtimes_{\alpha<\delta} P_\alpha$ let

$$\mathrm{supt}(p) = \{\alpha < \delta \mid p(\alpha) \neq \mathbb{I}_{P_\alpha}\},$$

the *support* of p, and for cardinal κ form the κ-product of $\langle P_\alpha \mid \alpha < \delta \rangle$ as

$$\{p \in \bigtimes_{\alpha<\delta} P_\alpha \mid |\mathrm{supt}(p)| < \kappa\}$$

ordered by

$$p \preccurlyeq q \Leftrightarrow \forall \alpha < \delta(p(\alpha) \preccurlyeq_{P_\alpha} q(\alpha))$$

(this all in M).

Show that the partial order P_κ used to add κ Cohen reals is the ω-product of κ copies of P_0, the partial order used to add one Cohen real.

[Easton [Eas70] gave a more general product, and used it to show that for regular cardinals κ the constraints given by monotonicity and by König's lemma 7.4.6(5) are the only constraints on the power function 2^κ.]

(7) **General iterated forcing.** Suppose that P is a partial order, and that $\mathring{Q}$ is a P-name for which $\mathbb{1}_P \Vdash$ "$\mathring{Q}$ is a partial order". Then we can consider the partial order

$$P * \mathring{Q} = \{\langle p, q\rangle \mid p \in P \wedge p \Vdash \mathring{q} \in \mathring{Q}\},$$

ordered by

$$\langle p, q \rangle \preccurlyeq \langle p', q' \rangle \Leftrightarrow p \preccurlyeq_P p' \wedge p \Vdash \mathring{q} \preccurlyeq_{\mathring{Q}} \mathring{q}'.$$

First show that we can restrict this to be a set in M; that is, there is a dense sub-ordering which is a set in M.

[This is one place where it is useful to restrict the terms used as names in the forcing language; essentially it is the redundant names, which will not denote anything new, which can make $P * \mathring{Q}$ into a proper class of M.]

Assuming this has been done, show that if G_1 is P-generic over M, and G_2 is Q^{G_1}-generic over $M[G_1]$, and we put

$$G_1 * G_2 = \{\langle p, q \rangle \in P * \mathring{Q} \mid p \in G_1 \wedge q^{G_1} \in G_2\},$$

then $G_1 * G_2$ is $P * \mathring{Q}$-generic over M and $M[G_1 * G_2] = M[G_1][G_2]$.

Conversely, show that if G is $P * \mathring{Q}$-generic over M and

$$G_1 = \{p \in P \mid \exists q (\langle p, q\rangle \in G)\}, \text{ and } G_2 = \{q^{G_1} \mid \exists p (\langle p, q\rangle \in G)\},$$

then G_1 is P-generic over M, G_2 is Q^{G_1}-generic over $M[G_1]$, and $G = G_1 * G_2$.

Show that the product in (5) is the special case when $\mathring{Q}$ is $\check{Q}$ for some $Q \in M$, and then $P \times Q$ is isomorphic to a dense subset of $P * \mathring{Q}$.

[Iterations of this form of product give the general iterations that are amongst the most powerful methods of forcing. See [ST71], [MS70], [She82].]

(8) **Martin's axiom.** If P is a partial order, and $\mathcal{D}$ is any collection of dense subsets of P, say that $G \subseteq P$ is $\mathcal{D}$-generic if G is consistent and upward-closed, and meets every member of $\mathcal{D}$.

Then *Martin's axiom for* κ, MA(κ), is:

Whenever P is a partial ordering with c.c.c., and $\mathcal{D}$ is a collection of at most κ dense subsets, then there is a $G \subseteq P$ which is $\mathcal{D}$-generic.

Show that MA(ω), Martin's axiom for ω, is provable in ZFC, and that MA(2^ω) is inconsistent.

[The point here is that we are, in effect, asking for generic subsets which are already in the ground model. Considering the forcing for adding one

Cohen real, with the dense sets used to show that the Cohen real was not in the ground model, gives the inconsistency for $\mathrm{MA}(2^\omega)$.]

Martin's axiom is then $\forall \kappa < 2^\omega (\mathrm{MA}(\kappa))$.

Show that the continuum hypothesis implies Martin's axiom, and that Martin's axiom without the restriction to c.c.c. is inconsistent with the negation of the continuum hypothesis.

[CH reduces Martin's axiom to $\mathrm{MA}(\omega)$. $\mathrm{MA}(\omega_1)$ is inconsistent without the restriction to c.c.c. by considering the partial order P_{ω,ω_1} of 8.10.9, with the dense sets used to show that the resulting function is onto ω_1.

Solovay and Tennenbaum [ST71] gave a model using iterated forcing which showed the consistency of Martin's axiom with the negation of the continuum hypothesis. Since then, many uses have been found for Martin's axiom, see e.g. [Frem84], [Kun80].]

8.12 Models for the independence of the axiom of choice

All of the forcing models presented so far satisfy the axiom of choice. Cohen's first presentation also gave models in which the axiom of choice failed, and we present a version of this now.

The method is to form models N intermediate between M and $M[G]$, which will contain only some of the new elements definable from G, but not G itself. The new elements used to give N will be those which are *symmetric*, in a sense which we shall describe. There will be enough symmetric elements to show that N is still a model of the ZF axioms, but the choice sets required for the axiom of choice (or the well-orderings required for the well-ordering theorem) will not be symmetric.

The symmetry will be determined by considering automorphisms of the set P of forcing conditions. In all the cases we shall consider in this section, we shall be using conditions which involve ω, and any permutation of ω can be considered as giving a permutation of the set P which respects the ordering of P; these will be automorphisms of P when P is considered as a partial ordering $\langle P, \preccurlyeq \rangle$, as it should be in all this chapter. To determine symmetry, we shall assume that we have some group $\mathcal{G} \in M$ of automorphisms of P, and also that we have selected certain subgroups of $\mathcal{G}$ which we consider to be *large*, and an element will be considered symmetric if the automorphisms which leave it fixed form a large subgroup in this sense. As in 5.3.13, the notion of a *filter* gives a suitable notion of largeness.

So we shall assume given a *normal filter of subgroups* $\mathcal{F} \in M$, which will be a non-empty set of subgroups of $\mathcal{G}$ with the following properties:

(i) if $H \in \mathcal{F}$ and $K \in \mathcal{F}$ then $H \cap K \in \mathcal{F}$;
(ii) if $H \in \mathcal{F}$ and $H \subseteq K$ where K is a subgroup of $\mathcal{G}$ then $K \in \mathcal{F}$;

(iii) if $\pi \in \mathcal{G}$ and $H \in \mathcal{F}$ then $\pi H \pi^{-1} \in \mathcal{F}$.

(Clause (iii) is the clause which makes the filter *normal*; we shall see why it is needed later.)

Now suppose $\pi : P \to P$ is in $\mathcal{G}$; we extend π to an automorphism of the names (or terms) used to construct $M[G]$, by the following definition:

Definition 8.12.1

$$\pi x = \{\langle \pi p, \pi y \rangle \mid \langle p, y \rangle \in x\}$$

for any $x \in M$. (Note that we are writing πp rather than $\pi(p)$ for the application of π to p; this reduces the number of parentheses, and we shall put them back if needed for clarity.)

It might make good sense to restrict attention from the start to those sets in M which are in fact relations, with domain a subset of P, since the model $M[G]$, and the submodel we are about to define, never make real use of any other members of M. The point is that if $x \in M$ has members which are not pairs $\langle p, y \rangle$ with $p \in P$, these are ignored in forming x^G and in all uses of $\overset{\circ}{x}$. However, it does not seem to make any of the work substantially clearer, and we could continue to consider every member $x \in M$ as we have done up to now. But the next definition will in fact make this restriction for us.

Given this definition of πx, there will be some sets x for which $\pi x = x$; such x are said to be *fixed* by π. For each $x \in M$ we shall define the *stabilizer subgroup* or *symmetry group* of x by

$$\text{sym}(x) = \{\pi \in \mathcal{G} \mid \pi x = x\}.$$

It is easy to see that this will always be a subgroup of $\mathcal{G}$.

Now the elements which we shall use to construct N will be those $x \in M$ for which $\text{sym}(x) \in \mathcal{F}$, and whose members all have the same property, and so on. We say that $x \in M$ is *hereditarily $\mathcal{F}$-symmetric* (or just *hereditarily symmetric*, when $\mathcal{F}$ is assumed known), and write $x \in \text{hsym}(\mathcal{F})$, if

$$\text{sym}(x) \in \mathcal{F} \text{ and, whenever } \langle p, y \rangle \in x, \ y \in \text{hsym}(\mathcal{F}).$$

This can be taken as a definition by induction over $\text{rank}(x)$, and will be well-defined and absolute over M whenever $\mathcal{F} \in M$. But note that $\text{hsym}(\mathcal{F})$ will be a proper class in the sense of M, so it is the formula $x \in \text{hsym}(\mathcal{F})$ which is absolute.

Now we define N, the *symmetric model* given by $\mathcal{F}$, as:

$$N = \{x^G \mid x \in M \wedge x \in \text{hsym}(\mathcal{F})\}$$

CONSTRUCTIBLE SETS AND FORCING

where G is P-generic over M.

Note that if $H = \mathrm{sym}(x)$ then $\pi H \pi^{-1} = \mathrm{sym}(\pi x)$. This explains why *normal* filters of subgroups are required: it shows that each automorphism in $\mathcal{G}$ will extend naturally to a permutation of N.

We can now state the main result of this section:

Theorem 8.12.2 *For any normal filter $\mathcal{F} \in M$ of subgroups of the automorphism group of P, and any G which is P-generic over M, the symmetric model N given by $\mathcal{F}$ is a model of ZF with $M \subseteq N \subseteq M[G]$.*

Towards proving this, let us first see that the names used to show that $M \subseteq M[G]$ will all be hereditarily symmetric. We have

$$\check{x} = \{\langle p, \check{y}\rangle \mid p \in P \land y \in x\},$$

from 8.8.9, so if we assume for induction that $\pi \check{y} = \check{y}$ for y with $\mathrm{rank}(y) < \mathrm{rank}(x)$ and any $\pi \in \mathcal{G}$, we shall have

$$\pi \check{x} = \{\langle \pi p, \pi \check{y}\rangle \mid p \in P \land y \in x\} = \{\langle p, \check{y}\rangle \mid \pi^{-1} p \in P \land y \in x\} = \check{x}$$

since π is an automorphism of P.

So the symmetry group $\mathrm{sym}(\check{x}) = \mathcal{G}$ for all these $\check{x}$, and we must have $\mathcal{G} \in \mathcal{F}$ for any normal filter of subgroups, since $\mathcal{F} \neq \emptyset$. So these names are all hereditarily symmetric, for any $\mathcal{F}$. Hence we shall always have $M \subseteq N$.

It is also easy to see that only the identity automorphism will preserve the name $\check{G}$; we leave that as an exercise. So unless $\mathcal{F}$ is trivial, and contains every subgroup of $\mathcal{G}$, we shall have $G \notin N$.

Most of the work in proving that N is a model of ZF, and in determining its other properties, can be summed up in appropriate versions of the definability and truth lemmas. A strong forcing relation can be defined for this symmetric model situation exactly as for the full model $M[G]$. The definition simply restricts the definitions in 8.11.1 and 8.11.3 to terms given by members of $\mathrm{hsym}(\mathcal{F})$, instead of by any members of M. So there is nothing new in proving the definability lemma. The proof of the truth lemma is also very little changed: we simply take all hereditarily symmetric members of M instead of all members of M, at appropriate places.

But in fact we need more that the truth lemma in this context; we also need the following:

Lemma 8.12.3 *For any $p \in P$ and any formula $\varphi(v_1, \ldots, v_k)$ of the forcing language $\mathcal{L}_M$ and for any automorphism π of P, we have*

$$p \Vdash^* \varphi(\mathring{x}_1, \ldots, \mathring{x}_k) \text{ if and only if } \pi p \Vdash^* \varphi(\pi \mathring{x}_1, \ldots, \pi \mathring{x}_k).$$

Proof The proof is by induction on the formula φ, and as usual most of the work is in the base case when φ is atomic, and that has to be proved by induction.

So suppose that $p \Vdash^* \overset{\circ}{y} \in \overset{\circ}{x}$; then from the definition we have some $\langle r, u \rangle \in x$ with $p \preccurlyeq r$ and $p \Vdash^* \overset{\circ}{u} = \overset{\circ}{y}$. Suppose from the induction hypothesis that $\pi p \Vdash^* \pi \overset{\circ}{u} = \pi \overset{\circ}{y}$. We know that $\langle \pi r, \pi u \rangle \in \pi x$ from the definition of πx, and we have $\pi p \preccurlyeq \pi r$ since π is an automorphism. So we shall have $\pi p \Vdash^* \pi \overset{\circ}{y} \in \pi \overset{\circ}{x}$. The converse follows by considering π^{-1}.

For the case of $p \Vdash^* \overset{\circ}{y} = \overset{\circ}{x}$, first note that if $Q \subseteq P$ is dense below p, then $\pi(Q) = \{\pi q \mid q \in Q\}$ will be dense below πp. So the various clauses in the definition in 8.11.1 for $=$ will all hold after the application of π or of π^{-1} if and only if they held before, just as for the case of $\in$ above. There is however one awkward point arising from the way we gave the definitions, and that is that it is conceivable that we might have $\text{rank}(z) < \text{rank}(x)$ but $\text{rank}(\pi z) > \text{rank}(\pi x)$; after all, we have not made any assumption about the ranks of the conditions $p \in P$, and we may have $\text{rank}(\pi p)$ very much greater than $\text{rank}(p)$ for some condition with $\langle p, u \rangle \in z$. The simplest way to see that this does not affect the truth of the lemma is to recast all the definitions and inductions in terms of a new rank function which reflects what we really need, instead of relying on the normal rank function which, while always to hand, may not be respected by the automorphisms in $\mathcal{G}$. One possible way is in the exercises.

The remaining cases of the proof of the lemma are also left as exercises.

8.12.4 The proof of 8.12.2

This is now completed by taking the terms $\overset{\circ}{t}$ which were defined when we proved the various axioms for $M[G]$ and showing that, if an automorphism π leaves the parameters unchanged, then we also have $\pi t = t$. Hence the symmetry group of t will contain the intersection of the symmetry groups of the parameters. So if these are all in $\mathcal{F}$, and there are only finitely many parameters in each case, the symmetry group of t must also be in $\mathcal{F}$. It will follow that if the parameters are hereditarily symmetric, so will be t and we shall have $t^G \in N$. So the axiom will hold in N, just as in the case of $M[G]$.

We give two examples, and leave the rest as exercises. First the pair-set: we showed in 8.9.1 that $\text{pair}(x, y) = P \times \{x\} \cup P \times \{y\}$ is a term which gives the unordered pair $\{x^G, y^G\}$. Suppose that x and y are both in $\text{hsym}(\mathcal{F})$, say $\text{sym}(x) = H \in \mathcal{F}$ and $\text{sym}(y) = K \in \mathcal{F}$. Let $\pi \in H \cap K$, then

$$\pi(\text{pair}(x, y)) = \pi(P) \times \{\pi x\} \cup \pi(P) \times \{\pi y\} = P \times \{x\} \cup P \times \{y\}$$

since $\pi \in H \Rightarrow \pi x = x$ and $\pi \in K \Rightarrow \pi y = y$, and $\pi(P) = P$ for any automorphism π of P.

CONSTRUCTIBLE SETS AND FORCING   183

So $H \cap K \subseteq \operatorname{sym}(\pi(\operatorname{pair}(x,y)))$ and $H \cap K \in \mathcal{F}$, so $\operatorname{pair}(x,y)$ is hereditarily symmetric, and the pair-set axiom will hold in N.

Now let us consider the replacement axiom. Comparing the proof in 8.9.5, we shall now suppose that

$$N \models \psi(y^G, z^G) \wedge \psi(y^G, w^G) \Rightarrow z^G = w^G$$

and by the truth lemma deduce

$$p \Vdash \psi(\mathring{y}, \mathring{z}) \wedge \psi(\mathring{y}, \mathring{w}) \Rightarrow \mathring{z} = \mathring{w} \text{ for some } p \in G \qquad (*)$$

as before. If a^G is the set whose image we require, we shall have $a \in \operatorname{hsym}(\mathcal{F})$; and we form Z_a and b as before, except that to deal with the awkward possibilities about ranks, we this time let

$$b = \{\langle q, z\rangle \mid \exists \langle q, y\rangle \in Z_a (q \Vdash \psi(\mathring{y}, \mathring{z}) \wedge \operatorname{rank}(z) < \beta)\}^M,$$

where β will be chosen later.

Now, provided β is at least as large as the ranks $\rho(q, y)$ used previously, the only point needed is to show that this b is hereditarily symmetric. So suppose $\pi \in \operatorname{sym}(a)$; we show that $\pi b = b$. (If there are parameters in ψ which are constant terms denoting further members of N, say $\mathring{x}_1, \ldots, \mathring{x}_k$ then we must take $\pi \in \operatorname{sym}(a) \cap \bigcap_{i \leq k} \operatorname{sym}(x_i)$.)

Suppose $\langle q, z\rangle \in b$; say $q \Vdash \psi(\mathring{y}, \mathring{z})$ with $\langle q, y\rangle \in Z_a$. Then we shall have $\pi q \Vdash \psi(\mathring{\pi y}, \mathring{\pi z})$, and we shall have $\langle \pi q, \pi y\rangle \in Z_a$ since if $\langle r, y\rangle \in a$ with $q \preccurlyeq r$ then we shall have $\langle \pi r, \pi y\rangle \in a$ with $\pi q \preccurlyeq \pi r$. Hence $\langle \pi q, \pi z\rangle \in b$ provided $\operatorname{rank}(\pi z) < \beta$. So if we take β large enough so that $\operatorname{rank}(\pi z) < \beta$ whenever $\operatorname{rank}(z) < \beta$, we shall have $\pi b = b$. Since there are only a set (in the sense of M) of permutations π, we can extend any ordinal in M to a larger ordinal also in M with this closure property; so there will be some $\beta \in M$ as required, and the replacement axiom will hold in N.

As noted, we leave the further details as exercises.

8.12.5   Exercises

(1) Show that the terms used in 8.9.2 and 8.9.3 to give the sum-set and power-set axioms will be hereditarily symmetric, if the parameter x is.
    Hence complete the proof that N is a model of ZF.
    Show also that the set f used in 8.9.6 for the axiom of choice cannot be expected to be hereditarily symmetric in general.
(2) Define a notion of rank_P as follows:

$$\operatorname{rank}_P(x) = \sup\{\operatorname{rank}_P(y) + 1 \mid \langle q, y\rangle \in x \text{ for some } q \in P\}.$$

Show that $\mathrm{rank}_P(x) \leq \mathrm{rank}(x)$ for every $x \in M$, and that all the definitions which used induction on the rank of terms can use induction on the rank_P instead.

However show that the collection $\{x \in M \mid \mathrm{rank}_P(x) < \alpha\}$ (for any $\alpha > 0$) will be a proper class of M, and we must restrict to sets which are relations with domain a subset of P if we are to avoid this.

Show that any hereditarily symmetric $x \in M$ must be a relation with domain a subset of P.

[Otherwise we never have $\pi x = x$ for any permutation π of P.]

Hence compute the ordinal β needed in the proof of 8.12.4 if this notion $\mathrm{rank}_P(x)$ is used in place of $\mathrm{rank}(x)$.

8.13 A model with a Dedekind-finite set

We give here one of the simplest cases where we can show that the axiom of choice fails. We shall take P as the poset P_ω as defined in 8.10.3, that is, the set of all finite functions from $\omega \times \omega$ into 2, ordered by extensions. (So this adds ω Cohen reals, if we take the full $M[G]$.)

Now suppose that $\pi : \omega \to \omega$ is a bijection, i.e. π is a permutation of ω. We consider π as an automorphism of P_ω by setting

$$\pi p(\pi n, m) = p(n, m).$$

It is easy to check that this is a bijection from P_ω onto itself, and that $\pi p \preccurlyeq \pi q$ if and only if $p \preccurlyeq q$. So it is an automorphism of P_ω. We let $\mathcal{G}$ be the set of all such automorphisms, arising from all permutations of ω, in the model M. (This will not be absolute; it is a nice exercise to check that $\mathcal{G}$ has the size of the continuum. But it will always be a set, so we shall have its relativization to M, and that is what we are taking here.)

Given any subset x of ω, some permutations will leave every member of that subset unmoved, i.e. $\pi n = n$ for each $n \in x$. Let $\mathrm{fix}(x)$ be the set of all such permutations; if we identify the permutations with the automorphisms of P_ω as above, it is easy to see that $\mathrm{fix}(x)$ will be a subgroup of $\mathcal{G}$. Now we take $\mathcal{F}$ as the filter of subgroups generated by the set of all $\mathrm{fix}(x)$ for x a *finite* subset of ω. This will be normal, since if $H = \mathrm{fix}(x)$ then it is easy to check that $\pi H \pi^{-1} = \mathrm{fix}(\pi''x)$. We form the model N as the symmetric model given by $\mathcal{F}$ and any G which is P_ω-generic over M.

Note that it will be straightforward in this context to check whether a term $\overset{\circ}{t}$ is symmetric: if we can find a finite subset of ω, say y, such that whenever π leaves every element of y fixed, then π leaves t fixed, then $\overset{\circ}{t}$ will be symmetric. We shall have to check the members of t, and so on, if we want to know that $\overset{\circ}{t}$ is hereditarily symmetric, of course.

CONSTRUCTIBLE SETS AND FORCING

Now we want to show that N contains a Dedekind-finite set, that is, a set A which is not finite but such that there is no embedding of ω into A.

The sets we want are the sets $x_n = \{m \in \omega \mid \exists p \in G(p(n,m) = 1)\}$, for $n \in \omega$ (which we recognize as Cohen reals), and then $A = \{x_n \mid n \in \omega\}$. But we cannot define them this way in N, since we shall not have $G \in N$; we must find suitable terms which will give these sets and which are hereditarily symmetric.

So consider the terms $\overset{\circ}{t}_n$, for fixed $n \in \omega$, given by

$$t_n = \{\langle p, \check{m}\rangle \mid p(n,m) = 1\}$$

where $\check{m}$ gives m as a member of N; and the term $\overset{\circ}{a}$ given by

$$a = \{\langle p, t_n\rangle \mid p \in P,\ n \in \omega\}.$$

These will give the sets required, since it is clear that $p \Vdash \check{m} \in \overset{\circ}{t}_n$ if and only if $\langle p, \check{m}\rangle \in t_n$, if and only if $p(n,m) = 1$, so that $t_n^G = x_n$; and every condition will force $\overset{\circ}{t}_n \in \overset{\circ}{a}$ for every $n \in \omega$. So we need to show these are hereditarily symmetric.

Consider any $\pi \in \mathcal{G}$, then

$$\begin{aligned}\pi t_n &= \{\langle \pi p, \pi(\check{m})\rangle \mid p(n,m) = 1\} \\ &= \{\langle \pi p, \check{m}\rangle \mid \pi p(\pi n, m) = 1\} \text{ since } \pi(\check{m}) = \check{m} \\ &= t_{\pi n}\end{aligned}$$

so that $\operatorname{sym}(t_n) = \operatorname{fix}(\{n\})$ and since all $\check{m}$ are hereditarily symmetric, so are all $\overset{\circ}{t}_n$ for $n \in \omega$. And

$$\pi a = \{\langle \pi p, \pi t_n\rangle \mid p \in P, n \in \omega\} = \{\langle \pi p, t_{\pi n}\rangle \mid p \in P, n \in \omega\} = a$$

since $\pi(P) = P$ and $\pi(\omega) = \omega$ for any $\pi \in \mathcal{G}$.

So the sets we want will be in N, given by $x_n = t_n^G$ and $A = a^G$. It is easy to see that A is infinite, since all the sets x_n will be distinct. To show that A is Dedekind-finite, we shall show that there can be no injection from ω into A.

So suppose that $\overset{\circ}{f}$ is a hereditarily symmetric term; we shall show that no condition p can force f^G to be an injection from ω into A. Suppose that $x \subseteq \omega$ is a finite set such that $\operatorname{fix}(x) \subseteq \operatorname{sym}(f)$, and that for contradiction, some condition $p \Vdash ``\overset{\circ}{f} : \check{\omega} \to \overset{\circ}{a}$ and $\overset{\circ}{f}$ is an injection". Since f^G is to be an injection, its range must be infinite, and we must have some $i, j \in \omega$ and some condition $q \preccurlyeq p$ such that $q \Vdash \overset{\circ}{f}(\check{i}) = \overset{\circ}{t}_j$, with $j \notin x$ and no pair $\langle j, m\rangle \in \operatorname{dom}(p)$ for any $m \in \omega$ (using the fact that x and $\operatorname{dom}(p)$ are finite).

Now we find a permutation π of ω in fix(x) (so that $\pi f = f$), and which fixes all $k \in \omega$ such that $\langle k, m \rangle \in \text{dom}(p)$ for any m (so we shall have $\pi p = p$); and also such that q and πq are compatible, but which moves j (so $\pi j \neq j$).

To find π, we find $j' \in \omega$, $j' \neq j$, which does not appear with $\langle j', m \rangle \in \text{dom}(q)$ for any m (and hence not in dom(p) either), and which is not in x, and let π interchange j and j' (and leave everything else fixed). Then if $\langle j, m \rangle \in \text{dom}(q)$, we have $\langle j', m \rangle \in \text{dom}(\pi q)$; but no $\langle j', m \rangle \in \text{dom}(q)$ and no $\langle j, m \rangle \in \text{dom}(\pi q)$ and so q and πq will be compatible.

But we have assumed $q \Vdash \overset{\circ}{f}(\check{\imath}) = \overset{\circ}{t}_j$, so we get $\pi q \Vdash \pi \overset{\circ}{f}(\check{\imath}) = \overset{\circ}{t}_{\pi j}$ which is

$$\pi q \Vdash \overset{\circ}{f}(\check{\imath}) = \overset{\circ}{t}_{j'}$$

and hence if r is any extension of both q and πq we shall have

$$r \Vdash \overset{\circ}{f}(\check{\imath}) = \overset{\circ}{t}_j \text{ and } r \Vdash \overset{\circ}{f}(\check{\imath}) = \overset{\circ}{t}_{j'}$$

which is a contradiction to the assumption $p \Vdash$ "$\overset{\circ}{f} : \check{\omega} \to \overset{\circ}{a}$ and $\overset{\circ}{f}$ is an injection", since we know that no condition can force $\overset{\circ}{t}_j = \overset{\circ}{t}_{j'}$ with $j \neq j'$.

8.13.1 *Exercises*

(1) *Amorphous sets.* An extreme type of Dedekind-finite set occurs when a set B is *amorphous*, that is, every subset of B is either finite or co-finite. Show that the set A produced above is not amorphous.

[The subset $B_n = \{x \in A \mid n \in x\}$ is easily seen to be both infinite and co-infinite, for any fixed $n \in \omega$.]

Hence show $\mathcal{P}(A)$ is not Dedekind-finite.

(2) A model with an amorphous set. To get this, we can add ω Dedekind-finite sets, each with the properties of the previous A. If these copies are X_n for $n \in \omega$, say $X_n = \{x_{n,m} \mid m \in \omega\}$ (where $x_{n,m}$ are Cohen reals), then $A = \{X_n \mid n \in \omega\}$ will be amorphous.

Show that we can do this in the model given by the poset $P = P_{\omega \times \omega}$, that is, take conditions p as finite partial functions from $\omega \times \omega \times \omega$ into 2, ordered by extensions.

[We shall want $x_{n,m} = \{l \mid \exists p \in G(p(n,m,l) = 1\}$, and the problem is to give an appropriate group of automorphisms and filter of subgroups so that these and the sets X_n will be symmetric; then A will also be symmetric.]

Take permutations of $\omega \times \omega$ given as follows: let π^0 and $\{\pi_n \mid n \in \omega\}$ be any permutations of ω, and define $\pi : \omega \times \omega \to \omega \times \omega$ by $\pi(n,m) = (\pi^0 n, \pi_n m)$. Then let these π act on the conditions p by:

$$\pi p(\pi(n,m), l) = p(n, m, l)$$

and take $\mathcal{G}$ to be the set of all the automorphisms of $P_{\omega \times \omega}$ given in this way (in the ground model M, as usual). Then let $\mathcal{F}$ be the filter of subgroups generated by the subgroups fix(u) for u a finite subset of $\omega \times \omega$.

(The effect of these permutations is that if $\pi(n,m) = (\overline{n}, \overline{m})$, then for any $m' \in \omega$, we have $\pi(n, m') = (\overline{n}, \overline{m}')$ with the *same* $\overline{n}$.)

Now show that this $\mathcal{F}$ is normal, and that the sets $x_{n,m}$ are given by the terms $\overset{\circ}{t}_{n,m}$ where

$$t_{n,m} = \{\langle p, \check{l}\rangle \mid p \in P \wedge p(n,m,l) = 1\} \text{ (where } \check{l} \text{ gives } l \text{ as usual)}$$

and the sets X_n and A by the terms $\overset{\circ}{T}_n$ and $\overset{\circ}{a}$ where

$$T_n = \{\langle p, t_{n,m}\rangle \mid p \in P, \ m \in \omega\}, \quad a = \{\langle p, T_n\rangle \mid p \in P, \ n \in \omega\}.$$

Show that these will be hereditarily symmetric and will give the sets required.

Finally to show that the set A will be amorphous, suppose that $\overset{\circ}{b}$ is any hereditarily symmetric term such that for some $p \in P$, $p \Vdash \overset{\circ}{b} \subseteq \overset{\circ}{a}$; suppose that fix($u$) $\subseteq$ sym(b). Let $y = \{n \in \omega \mid \exists m \in \omega[(n,m) \in u \cup \text{dom}(p)]\}$, and suppose some condition $q \preccurlyeq p$ has $q \Vdash \overset{\circ}{T}_n \in \overset{\circ}{b}$ with $n \notin y$. Then show that for any other $\overline{n} \notin y$ and any condition $r \preccurlyeq q$ there will be a permutation $\pi \in \mathcal{G}$ which fixes p and $\overset{\circ}{b}$, and with $\pi^0 n = \overline{n}$, and such that the conditions r and πq are compatible, and deduce that $q \Vdash \overset{\circ}{T}_{\overline{n}} \in \overset{\circ}{b}$ for every $n \notin y$.

[We shall have $\pi q \Vdash \overset{\circ}{T}_{\overline{n}} \in \overset{\circ}{b}$ and hence $\forall r \preccurlyeq q \exists r' \preccurlyeq r(r' \Vdash \overset{\circ}{T}_{\overline{n}} \in \overset{\circ}{b})$ which implies $q \Vdash \overset{\circ}{T}_{\overline{n}} \in \overset{\circ}{b}$. Hence for any extension $q \preccurlyeq p$ which decides any of these statements, we either have, for every $n \notin y$, $q \Vdash \overset{\circ}{T}_n \notin \overset{\circ}{b}$, or for every $n \notin y$, $q \Vdash \overset{\circ}{T}_n \in \overset{\circ}{b}$; since y is finite, this shows that A must be amorphous.

To find the permutation π, let $k \in \omega$ be larger than any integer occurring in dom(r), and define π by:

$$\pi^0 n = \overline{n}, \quad \pi^0 \overline{n} = n, \quad \pi^0 n' = n' \text{ for } n' \neq n, \overline{n};$$
$$\pi_{\overline{n}} m = \pi_n m = m + k \text{ for } m \leq k;$$
$$\pi_{\overline{n}}(m+k) = \pi_n(m+k) = m \text{ for } m \leq k;$$
$$\pi_{\overline{n}} m = \pi_n m = m \text{ for } m > 2k; \text{ and}$$
$$\pi_{n'} m = m \text{ for all } m \text{ and all } n' \neq n, \overline{n}.$$

Since n and $\overline{n}$ do not occur in dom(p) or in u this will fix p and $\overset{\circ}{b}$; and k was chosen so that πq and r cannot conflict.]

(3) A model with a countable family of pairs, with no choice function. Cohen gave this model in [Coh63b]. We add a countable family of pairs of Dedekind-finite sets, $A = \{Y_n \mid n \in \omega\}$ with $Y_n = \{X_{n,0}, X_{n,1}\}$, with no uniform way of choosing between the two members, i.e. A has no choice function in the model.

Show that this can be done in the model given by the poset $P = P_{\omega \times 2 \times \omega}$. Conditions will now be finite partial functions from $\omega \times 2 \times \omega \times \omega$ to 2, and we shall have Cohen reals $x_{n,i,m}$ for $n, m \in \omega$ and $i \in 2$ and the sets $X_{n,i}$ will be $\{x_{n,i,m} \mid m \in \omega\}$.

The permutations π wanted here will be given by permutations $\pi_{n,i}$ of ω and permutations ε_n of 2, and then π is given by $\pi(n, i, m) = (n, \varepsilon_n i, \pi_{n,i} m)$. $\mathcal{G}$ will be the group of all these permutations, and $\mathcal{F}$ will be the filter generated by subgroups which fix finite subsets of $\omega \times 2 \times \omega$. So π will fix the sets Y_n; it will switch the members of Y_n or not according to ε_n; and it will permute the members of $X_{n,i}$ according to $\pi_{n,i}$.

Show that symmetric names $\mathring{t}_{n,i,m}, \mathring{T}_{n,i}, \mathring{y}_n$ and $\mathring{a}$ can now be given for the wanted sets $x_{n,i,m}, X_{n,i}, Y_n$ and A, and also a symmetric name for a function $g : \omega \to A$.

Then show that, in the resulting symmetric model, there cannot be a choice function for A; suppose that

$$p \Vdash \mathring{f} : \mathring{a} \to \bigcup \mathring{a} \wedge \forall n (\mathring{f}(\mathring{y}_n) \in \mathring{y}_n)$$

for a symmetric term $\mathring{f}$, and get a contradiction by showing that if $p' \preccurlyeq p$ is such that $p' \Vdash \mathring{f}(\mathring{y}_n) = \mathring{T}_{n,0}$ for suitable n, then there is a permutation π which swaps $\mathring{T}_{n,0}$ and $\mathring{T}_{n,1}$ but leaves $\mathring{f}$ unchanged, and such that $\pi p'$ and p' are compatible.]

8.13.2 Fraenkel–Mostowski models

Models showing the independence of the axiom of choice from weaker set theories than ZF were produced much earlier than Cohen's work. The simplest weakening is that which allows urelements, by weakening the axiom of extensionality, as in 3.2.3(1), to give a theory sometimes known as ZFA, and sometimes as Fraenkel–Mostowski or FM (after two of the developers).

Given a model of ZFA, with a set A of urelements, and a group of permutations $\mathcal{G}$ acting on A, these permutations can be extended to act on the universe of all sets by the inductive definition

$$\pi x = \{\pi y \mid y \in x\}.$$

Then, given a normal filter of subgroups of the group $\mathcal{G}$, we can restrict to the hereditarily symmetric elements of the model, and show that we again get a model of ZFA but where the axiom of choice may fail.

In fact all of the models presented in this section were first thought of in this context of theory ZFA, and then translated into forcing models for ZF as presented above. A good exposition of this work is [Jec73].

8.14 Boolean-valued models: another presentation

Soon after forcing was discovered, Solovay gave an alternative presentation which we introduce here. We shall not give full details or prove all the results, but just an introduction to the ideas.

We have sketched in 5.3.16(9) how to construct a complete Boolean algebra from any partially ordered set. The partially ordered sets which we used to construct forcing models will now be replaced by these complete Boolean algebras.

If $\mathbb{B}$ is a complete Boolean algebra, we form the Boolean-valued model $V^{\mathbb{B}}$ in stages corresponding to the stages of the cumulative types. At each stage, instead of taking all subsets of the previous stage, we take functions from the previous stages into $\mathbb{B}$. The idea is motivated by thinking of the two-element Boolean algebra $\{\mathbb{O}, \mathbb{I}\}$, which is a complete Boolean algebra, and thinking of a subset (of anything) as given by its characteristic function (assigning $\mathbb{I}$ to members and $\mathbb{O}$ to non-members).

Within the cumulative type structure, all statements are either true or false; thus the atomic statements $a \in b$ and $a = b$ are true or false, and others are built from these. The two values {false, true} can be correlated with the elements of the two-valued Boolean algebra $\{\mathbb{O}, \mathbb{I}\}$. Within $V^{\mathbb{B}}$, all statements are given a value which lies in the complete Boolean algebra $\mathbb{B}$. For atomic statements $a \in b$, if b is a function from some stage which includes a, into $\mathbb{B}$, then the Boolean-value of $a \in b$ should be at least $b(a)$.

But care must be taken to ensure that such statements as $(a = c \wedge c \in b) \Rightarrow a \in b$ come out true, i.e. with Boolean-value $\mathbb{I}$; this means that $b(a)$ may not be the whole value of $a \in b$. One could restrict to functions b which satisfy an appropriate version of extensionality: they should at least satisfy $([a = c]^{\mathbb{B}} \wedge b(c)) \leq b(a)$ whenever both a and c are in the domain of b, where we write $[a = c]^{\mathbb{B}}$ for the Boolean-value of $a = c$. But there is also the further question of this value $[a = c]^{\mathbb{B}}$; since we want to make the axiom of extensionality true, we must make this equal the value given to $\forall x(x \in a \Leftrightarrow x \in c)$. In fact this can all be done in a consistent way, without restricting the functions taken, and defining the Boolean-values of all atomic statements by a transfinite induction which follows alongside the definition of the stages themselves. We indicate some of the steps in the exercises; they correspond roughly to the main steps of the proof of the definability lemma.

A good exposition is in [Bel85].

For compound statements, the Boolean-values are given by taking the obvious correspondence of the Boolean operations $\wedge$, $\vee$ and $'$ with the propositional connectives $\wedge$, $\vee$ and $\neg$ (and we have followed common custom in using the same symbols for these all along, except that there seems to be no agreement on the symbol for complement or negation). For the quantifiers, $\forall$ corresponds to $\bigwedge$ and $\exists$ to $\bigvee$, and it is here that it is important that the Boolean algebra be complete, so that these are always defined.

Now the basic result is that for any complete Boolean algebra, every axiom of first-order logic and of ZFC will have Boolean-value $\mathbb{I}$ in $V^{\mathbb{B}}$. This can be proved in ZFC; that is, for each such axiom φ there is a proof in ZFC of $[\varphi]^{\mathbb{B}} = \mathbb{I}$. (These proofs have quite a lot in common with the proof of the truth lemma, and the proofs of the axioms in forcing models.)

It follows that every theorem of ZFC will have Boolean-value $\mathbb{I}$. But if $\mathbb{B}$ is constructed from a partial ordering with appropriate properties, such statements as (for example) the continuum hypothesis can be made to have Boolean-value other than $\mathbb{I}$ (often $\mathbb{O}$), and so cannot be theorems of ZFC.

Of course, all this always has to assume that ZFC is consistent, otherwise everything will be provable; we can only give *relative* independence results. But we have *not* made here any assumption about countable $\in$-models; and one advantage of these Boolean-valued models is that they can give meaning to the idea of adding a generic set to the universe. Prima facie, the universe should already contain all subsets of any set, and there cannot be any more to add; there cannot really be any generic subsets of a non-trivial partial order, if we actually have the universe of all sets (i.e. the full second-order version of the cumulative type structure; then we cannot consistently assume that any G meets *every* dense set). But we can always think of the Boolean-valued universe, and this can have more sets than the original, two-valued universe.

The two presentations can be brought together by considering homomorphisms of the complete Boolean algebra $\mathbb{B}$ into the two-element Boolean algebra. Provided we take a suitably *complete* homomorphism, we can use this to define a map from the $\mathbb{B}$-valued universe into a two-valued universe. Such homomorphisms correspond exactly to ultrafilters in $\mathbb{B}$; and the required completeness corresponds to the ultrafilter being generic as a subset of $\mathbb{B}$. The two-valued universe will contain this generic ultrafilter, which will have been added to the original universe.

Of course, it will only be possible to prove that the generic sets or ultrafilters exist with the right properties, if we assume that we start with a countable standard model of ZFC.

8.14.1 Exercises

(1) Defining $V^{\mathbb{B}}$. Suppose that $\mathbb{B}$ is a complete Boolean algebra, and use the

following recursive definition:

$$V_\alpha^{\mathbb{B}} = \{x \mid \mathrm{Fn}(x) \wedge \exists \beta < \alpha(\mathrm{dom}(x) \subseteq V_\beta^{\mathbb{B}}) \wedge \mathrm{range}(x) \subseteq \mathbb{B}\};$$
$$V^{\mathbb{B}} = \{x \mid \exists \alpha(x \in V_\alpha^{\mathbb{B}}\}.$$

Then define the $\mathbb{B}$-value of formulas of the language $\mathcal{L}_{V^{\mathbb{B}}}$ as follows (here we shall not distinguish between the elements of $V^{\mathbb{B}}$ and the constant symbols which denote them in the language):

(i) $[x \in y]^{\mathbb{B}} = \bigvee_{z \in \mathrm{dom}(y)}(y(z) \wedge [x = z]^{\mathbb{B}})$;
(ii) $[x = y]^{\mathbb{B}} = \bigwedge_{z \in \mathrm{dom}(x)}(x(z)' \vee [z \in y]^{\mathbb{B}}) \wedge \bigwedge_{z \in \mathrm{dom}(y)}(y(z)' \vee [z \in x]^{\mathbb{B}})$;
(iii) $[\neg \varphi]^{\mathbb{B}} = ([\varphi]^{\mathbb{B}})'$;
(iv) $[\varphi \vee \psi]^{\mathbb{B}} = [\varphi]^{\mathbb{B}} \vee [\psi]^{\mathbb{B}}$;
(v) $[\exists x \varphi(x)]^{\mathbb{B}} = \bigvee_{z \in V^{\mathbb{B}}}[\varphi(z)]^{\mathbb{B}}$.

First show that the relation $x \in \mathrm{dom}(y)$ is well-founded on $V^{\mathbb{B}}$.

Then show that $a = [\varphi]^{\mathbb{B}}$ for atomic formulas φ can be defined by recursion on this relation, as a formula, with just the variables a, φ, and $\mathbb{B}$ free. (We are thinking of φ and $\ulcorner \varphi \urcorner$ as the same thing here, so that formulas are sets and we blur some distinctions which should be clear when needed by now.)

Then show that for each compound formula φ of the language $\mathcal{L}_{V^{\mathbb{B}}}$ we can define a formula $a = [\varphi]^{\mathbb{B}}$; we cannot treat φ as a variable here, but the formula being defined must grow in complexity as φ grows more complex.

Now show that all the axioms of first-order logic with equality are $\mathbb{B}$-valid; i.e. that $[\varphi]^{\mathbb{B}} = \mathbb{I}$ for every such axiom.

[Some details: we can show $x(z) \leq [z \in x]^{\mathbb{B}}$ now. The elements for which $x(z) = [z \in x]^{\mathbb{B}}$ for all $z \in \mathrm{dom}(x)$ are called *extensional*, and the whole construction can be done purely with extensional elements. But the details are not significantly simpler.

The form the equality axioms take will include:

$$[x = y]^{\mathbb{B}} \wedge [y = z]^{\mathbb{B}} \leq [x = z]^{\mathbb{B}}, \text{ and } [x = y]^{\mathbb{B}} \wedge [\varphi(x)]^{\mathbb{B}} \leq [\varphi(y)]^{\mathbb{B}}.$$

Much of the work for first-order logic was known earlier from [RS53].]
(2) Show that the Boolean-values for bounded quantifiers are what one would expect:

$$[\exists u \in x \varphi(u)]^{\mathbb{B}} = \bigvee_{u \in \mathrm{dom}(x)} (x(u) \wedge [\varphi(u)]^{\mathbb{B}}),$$

and similarly for $[\forall u \in x \varphi(u)]^{\mathbb{B}}$.
(3) Complete subalgebras. If $\mathbb{B}' \subseteq \mathbb{B}$ is a subalgebra, it is said to be *complete* if for any $X \subseteq \mathbb{B}'$, $\bigvee X$ is the same whether computed in $\mathbb{B}$ or in $\mathbb{B}'$. (Note that *finite* meets and joins must be the same, wherever they are

computed, by definition of a subalgebra; but this is far from necessarily true for infinite meets and joins.)

Show that if $\mathbb{B}'$ is a complete subalgebra of $\mathbb{B}$ then also $\bigwedge X$ will be the same, whether computed in $\mathbb{B}$ or in $\mathbb{B}'$.

Show further that if $\mathbb{B}'$ is a complete subalgebra of $\mathbb{B}$, then $V^{\mathbb{B}'}$ will be a submodel of $V^{\mathbb{B}}$; and for formulas φ with only bounded quantifiers, and with parameters from $V^{\mathbb{B}'}$, $[\varphi]^{\mathbb{B}} = [\varphi]^{\mathbb{B}'}$.

The most important case of this is the two-element Boolean algebra $\mathbf{2} = \{\mathbb{O}, \mathbb{I}\}$, which is a complete subalgebra of every Boolean algebra. Show that $V^{\mathbf{2}}$ is isomorphic to V, so that we can consider $V^{\mathbb{B}}$ as an extension of V in a natural way.

[This embedding of V into $V^{\mathbb{B}}$ is very similar to the embedding of M into $M[G]$ in 8.8.9, or rather 8.11.5(1), since $\mathbb{I}$ is available.]

(4) Now show that for any complete Boolean algebra, $V^{\mathbb{B}}$ will be a model of ZFC.

[Most of the steps here are similar to the proofs given in 8.9 for the forcing case. We shall not give further details of the development in Boolean-valued terms, but refer the reader to e.g. [Bel85] or [Sco67].]

9

Miscellaneous further topics

9.1 Introducing variables for classes

On several occasions in discussing sets we have wanted to refer to collections of sets which were not themselves sets (i.e. proper classes), or were not at the time known to be sets. V, the universe of all sets, and On, the class of all ordinals, are probably the commonest proper classes. Many workers, probably going back to Cantor, have noted that it is not simply talking about such collections that causes paradoxes; paradoxes arise only when we assume that such collections could be members of other collections (or indeed of themselves), or that they could be arguments of functions (which for us is the same thing). Von Neumann seems to have been the first to work systematically in a setting which allowed arbitrary collections of sets as classes (some of which would be proper classes, others would be sets), with variables for classes, and with a restriction on the comprehension principle which acknowledges the distinction between sets and proper classes in order to prevent the known paradoxes. Since then many variants of such systems have been given, and we present two such.

9.2 System VNB

This was given by Bernays [Ber54] as a variant of von Neumann's system; it was then used by Gödel in his monograph on the consistency of the generalized continuum hypothesis, and is sometimes known by the acronym VNBG or even NBG.

9.2.1 Language

The language for this system has just one sort of variable, which we shall call class variables and write as $X, Y, Z, \ldots$

We then make the definition:

Definition 9.2.2 $\mathfrak{M}(X)$ for $\exists Y(X \in Y)$ (X is a set; $\mathfrak{M}$ is from the German "*Menge*", meaning "set").

$x, y, z, \ldots$ as restricted variables for sets; so $\forall x \varphi(x)$ abbreviates $\forall X(\mathfrak{M}(X) \Rightarrow \varphi(X))$, etc.

Note that this strict distinction between capital and lower-case letters for variables has *not* been used in the rest of this work; it will apply only for this section.

Now the distinctive feature of von Neumann's system was its axiom of comprehension. The leading idea is that since the proper classes are not simple objects—not given; in Cantor's phrase, not objects of mathematics (but rather our constructions), it may not make good sense to quantify over them; or at least we should not claim to know too much about the results of such quantification. In particular, we should not think we already have the collection of all proper classes; and we should not construct other classes using formulas which quantify over all proper classes. The contrast is with the sets, which are taken to be simple, given objects and which we can quantify over with assurance. This view is known as the *predicative* view of proper classes, and conforms with Russell's *vicious circle principle*: "objects cannot properly be defined by quantifying over collections which include those objects". In contrast to the sets, which are considered given (and we are trying to find out about them, so that definitions of sets pick out sets which already exist, and the vicious circle principle can properly be rejected for sets), the proper classes are to be thought of as collections which we are defining, which may not exist before we construct them. Hence the comprehension principle takes the restricted form:

9.2.3   Predicative comprehension principle

For every formula $\varphi(x)$ with no bound class variables (i.e. with all bound variables restricted to sets, as above), we have the axiom

$$\exists X \forall x(x \in X \Leftrightarrow \varphi(x)) \quad \text{i.e. we can form the class} \quad \{x \mid \varphi(x)\}.$$

This schema does allow *free* class variables in the formula $\varphi(x)$, which will then be parameters. So for example we can form $\mathcal{P}(X) = \{x \mid x \subseteq X\}$, the power class of any given class X, noting that we must write $x \subseteq X$ as $\forall y(y \in x \Rightarrow y \in X)$, using a bound *set* variable y; but this is natural since $y \in x$ implies y is a set anyway.

The remaining axioms are very similar to the Zermelo–Fraenkel axioms: the set-existence axioms (null-set, pair-set, sum-set, power-set, and axiom of infinity) can indeed be exactly the same, and we do not need to re-write them. But note that the pair-set, sum-set, and power-set axioms can now be read as: if the parameters involved (e.g. a, b) are *sets*, then the resulting class (e.g.

MISCELLANEOUS FURTHER TOPICS

$\{a, b\}$) is also a set. But that resulting class was already given as a *class* by the comprehension axiom, since all of these axioms will naturally use only bound set variables.

Extensionality and foundation will apply also to classes; indeed it is hard to think of collections as being taken as things in any sense without assuming that extensionality holds for them, and we already used this in presenting the axiom of comprehension, in claiming that the axiom allows us to form the class $\{x \mid \varphi(x)\}$ (since that implies introducing a defined term, which requires that we can prove uniqueness first as in 2.5, which requires extensionality for the class concerned). There is no way to prove extensionality for classes from extensionality for sets alone. But for the axiom of foundation, the class form follows from the set form, given the other axioms (see exercise 9.2.6(3)).

(i) Extensionality: $\forall x (x \in X \Leftrightarrow x \in Y) \Rightarrow X = Y$;
(ii) Foundation: $\exists x (x \in X) \Rightarrow \exists x (x \in X \wedge x \cap X = \emptyset)$.

The subset and replacement axiom schemes will change, in that they no longer need to use arbitrary formulas; instead, they can use free class variables and become single axioms:

(iii) Subset:
$$\exists x \forall y [y \in x \Leftrightarrow y \in a \wedge y \in X];$$
i.e. $\{y \mid y \in a \wedge y \in X\}$ or $a \cap X$ is a set for any class X (given that a is a set).

(iv) Replacement:
$$\forall z \forall u \forall v [\langle z, u \rangle \in X \wedge \langle z, v \rangle \in X \Rightarrow u = v]$$
$$\Rightarrow \exists x \forall y [y \in x \Leftrightarrow \exists z (z \in a \wedge \langle z, y \rangle \in X)];$$

i.e. if X is any functional class, then (given that a is a set), $\{y \mid \exists z (z \in a \wedge y = X(z)\}$ or $\{X(z) \mid z \in a\}$, the image of a under X, is a set.

The axiom of choice will also be left exactly as in Zermelo–Fraenkel. But it may be noted that von Neumann argued for a class form of the axiom of choice: he argued that every proper class (not just every set) should be well-ordered, and gave the axiom in the form:

(v) $V \sim \text{On}$, i.e. $\exists F (F : V \to \text{On} \wedge F$ is one–one onto).

This is equivalent to a global choice function (a choice function on V): $\exists F [F : V \to V \wedge \forall x \in V (F(x) \in x \vee x = \emptyset)]$. (See exercise 9.2.4(2).)

It is not easy to see any justification of this form of axiom of choice, simply from the cumulative hierarchy. It implies a uniformity in selecting choice functions (say for each rank of the cumulative hierarchy in turn) which does not seem to be justified unless we are thinking of taking all possible proper classes. But if we could take all possible proper classes, that collection would simply be the next stage of the cumulative hierarchy and the question arises, why have we stopped? If the cumulative hierarchy has been taken through all possible stages, we must not be able to form the collection of all possible proper classes; so why should we be able to form a uniform collection of choice functions?

However if there is some uniformity in the way in which each stage follows from the stages before, then the global axiom of choice is to be expected; and Gödel indeed showed that it follows from the axiom of constructibility (see 8.5).

9.2.4 A conservative extension result

We have set out the system VNB with one sort of variable; an equivalent way is to use two sorts of variables, one for sets $(x, y, z, \ldots)$ and the other for classes $(X, Y, Z, \ldots)$, with an axiom of extensionality between them (i.e. $\forall x(x \in Q \Leftrightarrow x \in Q') \Rightarrow Q = Q'$ where Q and Q' are either set or class variables, not necessarily the same). $\mathfrak{M}(X)$ can then be taken as $\exists x(x = X)$, and the other axioms can remain unchanged as written. Then we can regard VNB as an extension of ZF in a literal sense: every formula of ZF is a formula of VNB; every theorem of ZF is a theorem of VNB. But we have more; we also have:

Theorem 9.2.5 *VNB is a conservative extension of ZF; i.e. if φ is a formula of ZF, then if φ is a theorem of VNB, it is already a theorem of ZF.*

Indications of the proof are in exercise 9.2.6(4). We simply note for now that this means that we have not added anything new to set theory: if we can prove statements in VNB, and those statements are purely in the language of ZF (so not using class variables at all), then we could have proved the statement in ZF.

9.2.6 Exercises

(1) Show that the classes are closed under Boolean operations ∪, ∩, and complementation; and using the axiom of extensionality as given, that they form a Boolean algebra.
    [This is probably the closest we can come in this work to the original intentions of George Boole.]
(2) Show the equivalence of the two global forms of the axiom of choice.
    [From a global choice function F, use transfinite induction to define a function G from On to V which is onto. At each step, use F to choose an element of lowest rank as the next value of G. Von Neumann gave this construction when he introduced transfinite induction, [vN28b].]
(3) Show that the class form of the axiom of foundation follows from the set form.

MISCELLANEOUS FURTHER TOPICS   197

[Use the ranks; show that a disjoint member of $X \cap V_\alpha$ must be a disjoint member of X.]

(4) Show that any structure which is a model of ZF can be expanded to a structure which is a model for VNB. Hence deduce that VNB is a conservative extension of ZF.

[Here the *expansion* of a structure involves adding more parts to the structure. The structures considered in 2.4 were of the form (D, E), with D a domain and E a binary relation on that domain. We must now expand them to the form (C, D, E) where C is a new domain which will interpret the class variables; D will still interpret the set variables. Take C as the collection of all subsets of D which are definable with parameters from D, in other words the definable power-set of D as in 8.1. Then (C, D, E) will satisfy the predicative comprehension principle. Showing this for the classes X of the form $\{x \mid \varphi(x)\}$ where $\varphi(x)$ has free class variables will involve considering the assignment of those class variables first; if they are assigned to values in C, then we shall have corresponding formulas, which will need to be substituted carefully into $\varphi(x)$ to get a formula $\varphi'(x)$ with no class variables, which will give X.]

This is certainly not true for the next system, proposed by Morse [Mor65] and used earlier by Kelley [Kel55] and so known by the acronym MK.

9.3   System MK

This system simply rejects the restrictions of predicativity, and allows *any* formulas in the comprehension principle. So for any formula $\varphi(x)$ we can form the class $\{x \mid \varphi(x)\}$. The remaining axioms are taken as in VNB.

We give here just one example of the freedom which this allows: we can, in MK, define the ordinals in a way which is exactly parallel to the way we defined the natural numbers in 6.1:

Definition 9.3.1 $\text{Ind}^+(X)$ for

$$\emptyset \in X \wedge \forall x(x \in X \Rightarrow x \cup \{x\} \in X) \wedge \forall x(x \subseteq X \Rightarrow \bigcup x \in X),$$

(X is a *strongly inductive* class).

On for $\{x \mid \forall X(\text{Ind}^+(X) \Rightarrow x \in X\}$.

(Note the bound class variable in this definition of On.)

So a strongly inductive class is closed under successors (in the usual sense for von Neumann ordinals), and also under taking unions of subsets. Of course there are many classes besides On that have this property (e.g. V). On is then the smallest such class. We leave it as an exercise to prove that this agrees with the usual definition.

9.3.2 Other properties of MK

The proofs given above do not show the main fact that should be noted about the impredicative comprehension principle: MK is *not* a conservative extension of ZF. In fact one can show in MK that ZF is consistent. The usual way to show this requires the coding of syntactic and semantic notions within set theory, as in 8.1.

If we take the language $\mathcal{L}_V$, allowing constant symbols for all sets, and restrict to formulas with no free variables, then using a class variable we can give a definition of truth for such formulas; this will use just one class variable to denote the class of all true formulas. Just to say of a class X that it contains all the true formulas, can be done using bound set variables only; but as soon as we want to say that such a class X *exists*, we shall need a bound class variable and so MK (and VNB will not suffice). Once we have such a class, we shall be able to prove that ZF is consistent. For the codes for all the axioms of ZF will be in X, but the code for $0 = 1$, for example, will not be; and we shall be able to show that X is closed under the rules of deduction as given in chapter 2.

Of course, this will be a proof within MK; since MK extends ZF, if ZF is in fact inconsistent, then ZF (and MK) can prove everything, including the consistency of ZF. But if ZF is in fact consistent, this shows that MK can prove more than ZF and so is not a conservative extension.

9.3.3 Natural models

It is worth noting why the construction given in exercise 9.2.6(4) may break down when considering the system MK. First, there is no easy way to ensure that the classes C added when expanding the given structure (D, E) will satisfy the impredicative comprehension principle; the only way that will clearly satisfy this, is to add all possible subclasses of D. But then there is no reason to expect the class form of the replacement axiom or the foundation axiom to hold. If the structure (D, E) is not in fact well-founded, then some subclass of D will be a descending E-chain and will contradict the axiom of foundation. If there is in fact a mapping from a set in D which is cofinal in the ranks of D (which must of course not be definable in (D, E) since that satisfies the replacement axiom), then the class coding this mapping will contradict the axiom of replacement.

Note that the *Skolem paradox* rests on the observation that there must, if ZF is consistent, be a model (D, E) for ZF which is countable (i.e. D is countable); this follows from the Löwenheim–Skolem theorem, see [CK73]. (But note that we do not get a *transitive* countable model, in general.) Such a countable model of ZF would certainly allow a mapping from a set in D, namely the set representing ω in (D, E), to the whole of D; which illustrates

MISCELLANEOUS FURTHER TOPICS   199

the points noted above.

In fact if we add all possible subclasses as suggested, and we do get a model of MK, then we must have started with a model (D, E) of a particularly nice form, which Tarski called a *natural model* of ZF. We comment further on these in connection with inaccessible cardinals below.

9.3.4 Exercises

(1) Show that a strongly inductive class must contain all ordinals; and that the definition in 9.2.1 agrees with the definition in 6.5.
(2) Give a proof in MK of the theorem on transfinite recursion which "cuts down from above", parallel to the proof of the recursion theorem in exercise 6.2.15(2). [Given $F : V \to V$ we must find $H : \text{On} \to V$ such that for all ordinals α, $H(\alpha) = F(H \upharpoonright \alpha)$. What property of a class X will ensure that it contains H? Show that the following will work:

$$\forall \alpha \forall y [y \subseteq X \wedge \text{dom}\, y = \alpha \Rightarrow \langle \alpha, F(y) \rangle \in X].$$

Then take H as the intersection of all such X (using impredicative comprehension), and prove that it must be a function with domain On.]
(3) Give a definition of truth for formulas of ZF within MK.

[Use the apparatus developed in 8.1, but now replacing the set X in definitions 8.1.2 and 8.1.3 by the class V. Some adjustments are needed to take care of the fact that we cannot think of satisfaction sequences in quite the same way as before; first we must recast the notion of a finite sequence of classes as a single class, using a definition such as

$$S(i) = \{a \mid \langle i, a \rangle \in S\},$$

since we shall not get what we want if we form the pair $\langle i, S(i) \rangle$ if $S(i)$ is a proper class.

With suitable adjustments, show that we can define $\text{SatSeq}(S, x, V)$ and hence $\text{Sat}(V, \ulcorner \varphi \urcorner, a)$ and from that the set of formulas valid in $\langle V, \in \rangle$. Note that these will be formulas of ZF, with no class variables.

Tarski's paradox is the observation that no such definition of truth for formulas of a system can be possible within the same system, unless the system is inconsistent; see [Tar33].]

9.4 Axioms of extent

The only axioms of extent we considered in chapter 3 were the axioms of infinity and replacement. These suffice to prove the existence of cardinals $\aleph_\alpha$ for all ordinals α, and with the power-set axiom also the existence of the ranks

V_α. If the axiom of choice is used, we then can take the cardinals of the V_α as initial ordinals, usually denoted by $\beth_\alpha$. ($\beth$ is the second Hebrew letter, beth.) It was early noted that the only cardinals of the form $\aleph_\alpha$ or $\beth_\alpha$ with α a limit ordinal, which can be proved to exist in ZF, are *singular*, in that each is the limit (or union) of a smaller number of smaller cardinals (see exercise 7.4.6(2)). Cardinals of the form $\aleph_\alpha$ or $\beth_\alpha$ with α a limit ordinal are called *limit cardinals* or *strong limit cardinals* respectively, and no-one has proved that such cardinals (of either form) *must* be singular. From the point of view of the heuristic picture of the cumulative types, a strong limit cardinal $\beth_\alpha$ corresponds to a place where the collection of all sets existing at levels up to (but not including) that point, forms a model of all the axioms of ZFC except possibly the replacement axiom. In order to satisfy the replacement axiom also, we would most naturally think of the second-order form. So we would expect $\beth_\alpha$ to be *regular*, not singular (and we would also have to have $\beth_\alpha = \alpha$, see 7.4.6(2)). Such a cardinal $\beth_\alpha$ is called a *strongly inaccessible cardinal*, and then V_α was called by Tarski a *natural* model of ZFC; see 9.4.3(2).

9.4.1 Axioms for inaccessible cardinals

If we want to say that the cumulative hierarchy goes further than we can think of, then it is natural to assume the existence of many strongly inaccessible cardinals, since the axioms we have thought of (the axioms of ZFC) take us naturally to such points. A simple way to say that there are many inaccessible cardinals is

$$\forall \alpha \exists \kappa > \alpha (\text{Inac } \kappa)$$

where Inac κ is defined as Init $\kappa \wedge \text{Reg } \kappa \wedge (\mu < \kappa \Rightarrow 2^\mu < \kappa)$. These will be strongly inaccessible cardinals; a similar axiom could be given for weakly inaccessibles, which are regular limit cardinals (i.e. not necessarily strong limits). However it is hard to see this as finishing the question; we indicate in 9.4.3(3) how further natural questions arise.

9.4.2 Other large cardinal axioms

Many properties have been found which imply that some cardinal is inaccessible or strongly inaccessible, usually when investigating problems arising more or less naturally in analysis or set theory. One of the earliest and most important was Ulam's problem as to whether a measure could be both countably additive and defined on all subsets of an infinite set. This leads quickly to the notion of a *measurable cardinal*, and the first result shown was that such a cardinal must be strongly inaccessible. It was some years before it was shown that the first strongly inaccessible cardinal cannot be measurable; indeed the first measurable cardinal must be incomparably

MISCELLANEOUS FURTHER TOPICS

greater than the first strongly inaccessible, or any of the cardinals indicated in 9.4.3(3). Many other properties are now known which give large cardinals, and any assumption that such a large cardinal exists can be regarded as a further axiom of extent. The relations between them have been much studied, and one important use of them is as a scale to measure consistency strength: any first-order theory, if we assume it to be consistent, will define a certain strength of consistency when we ask, which other theories can we then prove consistent? Large cardinal axioms have been found to be the most useful measures for consistency strength, at levels above that of ZFC. See [Kan94] for a recent exposition.

9.4.3 Exercises

(1) Show that the cardinals $\beth_\alpha$ defined in 9.4 satisfy the recursive definition

$$\beth_0 = \aleph_0; \quad \beth_{\alpha+1} = 2^{\beth_\alpha}; \quad \text{and}$$

$$\beth_\lambda = \bigcup_{\alpha<\lambda} \beth_\alpha \text{ for limit } \lambda.$$

(2) Natural models of ZFC, VNB and MK. A structure of the form $\langle V_\alpha, \in \rangle$ which satisfies the axioms of ZFC is called a *natural model* of ZFC; a structure of the form $\langle V_{\alpha+1}, V_\alpha, \in \rangle$ which satisfies the axioms of VNB is called a natural model of VNB.

Show that $\langle V_{\alpha+1}, V_\alpha, \in \rangle$ is a natural model of VNB if and only if α is a strongly inaccessible cardinal; and in that case it is also a model of MK, and $\langle V_\alpha, \in \rangle$ is a natural model of ZFC.

[But it should be noted that if there is a natural model $\langle V_{\alpha+1}, V_\alpha, \in \rangle$ of VNB (i.e. if there is a strongly inaccessible cardinal α), then there will be natural models $\langle V_\beta, \in \rangle$ of ZFC which are smaller, i.e. with $\beta < \alpha$. The first such β will have cofinality ω and can be shown to exist by introducing a collection of Skolem functions (as in 8.6.1) for the structure $\langle V_\alpha, \in \rangle$ and finding the first ordinal β such that V_β is closed under them.]

(3) Hyperinaccessible cardinals. First, show that the axiom for inaccessible cardinals in 9.4.1 is equivalent to the existence of a proper class of strongly inaccessible cardinals.

Assuming this holds, let κ_α enumerate this proper class, and let θ be a fixed point of this enumeration (so that $\kappa_\theta = \theta$). Such a θ is then called *hyperinaccessible*. Show that this is equivalent to: θ is regular and has θ strongly inaccessible cardinals below it.

[Note that the enumeration used here will not be continuous at limits, i.e. we will not have $\kappa_\lambda = \bigcup_{\alpha<\lambda} \kappa_\alpha$ for all limits λ but in general $\bigcup_{\alpha<\lambda} \kappa_\alpha$ will be singular, and κ_λ will be the next inaccessible above it.]

Show also that if θ is hyperinaccessible, then $\langle V_\theta, \in \rangle$ will be a model of the axiom for inaccessible cardinals in which there are no hyperinaccessible cardinals.

[Within these natural models, notions such as regular cardinals and strong limit cardinals are easily seen to be absolute for the model in question, and this is all that is needed.]

9.5 Other presentations of set theory

Various alternative presentations of set theory have been given. Some aim to present the cumulative type structure in a clearer or simpler way, others present different conceptions altogether.

9.5.1 Axioms using ranks

First we give a presentation due to Montague and Scott ([Sco74]), which is arguably a much more natural way to present the axioms for the cumulative structure of types, particularly the axiom of foundation. They introduce first new variables for ranks, i.e. for the sets V_α of exercise 6.6.13(3); we shall use R, R', R'', etc. as variables for ranks. It is simplest to regard these as a new sort of variable, but with axioms to say that these are sets, and that extensionality holds both for these and for other sets (which may or may not be ranks). We take also the subset axiom in the usual form, and add just two further axioms for ranks:

9.5.2 Restriction

$$\forall x \exists R (x \subseteq R); \text{ and}$$

9.5.3 Accumulation

$$\forall R' \forall x [x \in R' \Leftrightarrow \exists R \in R' (x \in R \lor x \subseteq R)].$$

So restriction says that every set is a subset of some rank, and accumulation says that members of a rank are just those sets which are either members of an earlier rank (so that the ranks are cumulative), or are subsets of an earlier rank. So far these axioms are consistent with there being just one set, the empty set, which is also the empty rank.

Now an interesting point of these axioms is that the axiom of foundation follows from them, without any further assumptions. We can also prove that the ranks are linearly ordered (and hence well-ordered). For the remaining

MISCELLANEOUS FURTHER TOPICS

axioms, the sum-set axiom will follow (using the subset axiom; it will be a subset of any rank to which the given set belongs). The pair-set and power-set axioms will need some axiom of extent, to ensure that there is no last rank; and the axioms of infinity and replacement will need much stronger axioms of extent. These can be given as reflection principles very much in the spirit of the presentation in terms of ranks. We sketch these developments in exercises.

9.5.4 Exercise

(1) Show that $R \in R' \Rightarrow R \subseteq R'$ from the axiom of accumulation; and using Russell's paradox show that $R \notin R$.

[Form $a = \{x \in R \mid x \notin x\}$, and assume $R \in R$, and deduce Russell's paradox.]

Now use the paradox of groundedness (as in 1.5.1(1)) to deduce the axiom of foundation.

Let $\mathbf{grdd}\,x$ be $\forall a[x \in a \Rightarrow \exists y \in a(y \cap a = \emptyset)]$ (x is grounded, as in 1.5.1(1)), and let $\|R\|$ be $\{x \in R \mid \mathbf{grdd}\,x\}$. Show that

$$R \in R' \Rightarrow \|R\| \in \|R'\|.$$

[$\|R\| \in R'$ follows from the axiom of accumulation directly; to show that $\|R\|$ is grounded, suppose $\|R\| \in a$. Either $\|R\| \cap a = \emptyset$, or some $x \in \|R\| \cap a$; but then x is grounded and $x \in a$ so we get $y \in a$ with $y \cap a = \emptyset$. Either way we have shown $\|R\|$ is grounded.]

Now for any formula $\Phi(R)$ show

$$\exists R\Phi(R) \Rightarrow \exists R[\Phi(R) \wedge \neg \exists R'(R' \in R \wedge \Phi(R'))]. \qquad (*)$$

[Given $\Phi(R)$, take $a = \{b \mid \exists R' \in R[\Phi(R') \wedge b = \|R'\|]\}$. If a is empty we are done; R already has the required property. But if $a \neq \emptyset$ then since every member of a is grounded, we get some $b \in a$ with $b \cap a = \emptyset$; and then this b is of the form $\|R'\|$ for some R' with the property required.]

From this deduce

$$R \in R' \vee R = R' \vee R' \in R.$$

in other words the ranks are linearly ordered.

[Suppose not, and in $(*)$ take $\Phi_0(R)$ as $\exists R' \neg (R \in R' \vee R = R' \vee R' \in R)$ and suppose R_0 is minimal for this; then take $\Phi_1(R)$ as $\neg(R_0 \in R \vee R_0 = R \vee R \in R_0)$, and suppose R_1 is minimal for *this*. Now if $R \in R_0$, then $\neg \Phi_0(R)$ and hence $R \in R_1 \vee R = R_1 \vee R_1 \in R$. Since $R \in R_0$ we cannot have $R = R_1 \vee R_1 \in R$, so $R \in R_0 \Rightarrow R \in R_1$. And similarly if $R \in R_1$ then $\neg \Phi_1(R)$, so $R_0 \in R \vee R_0 = R \vee R \in R_0$. Again $R_0 \in R \vee R_0 = R$ will be impossible, and we have shown $R \in R_0 \Leftrightarrow R \in R_1$. But this gives $R_0 = R_1$ by extensionality and accumulation, a contradiction.]

Further development of this sort of axiom system is in [Sco74]; in particular an appropriate form for the axioms of infinity and replacement is given in the form of a reflection principle.

9.5.5 Quine's set theory

Willard van Orman Quine [Qui37] gave a presentation of the foundations of mathematics which was a fundamentally different reaction to the paradoxes. His basic observation was that all the known paradoxes used comprehension for *unstratified* formulas (as defined below), while the axioms he wanted could all be given by comprehension for *stratified* formulas. He worked in a language with just $\in$ and with equality defined by extensionality, and for formulas of this language the distinction is the following:

Definition 9.5.6 A formula φ is *stratified* if each variable of φ can be labelled with an integer in such a way that for every atomic subformula of φ of the form $x \in y$, the label of y is one greater than the label of x.

Note that this implies that in a stratified formula including a subformula of the form $x = y$, x and y must get the same label. The idea can be seen as a sort of minimal residue of the Russell notion of types (non-cumulative); the resulting sets are not labelled with a type, but the only allowable references to sets (within abstraction terms) must be as if the sets did have strict types.

Quine proposed just two axioms: the axiom of extensionality in the form

$$\forall z(z \in x \Leftrightarrow z \in y) \Rightarrow \forall u(x \in u \Leftrightarrow y \in u)$$

(this is equivalent to the usual form of the axiom of extensionality when equality is defined instead of taken as primitive), and the stratified comprehension principle

$$\exists x \forall y(y \in x \Leftrightarrow \varphi(y))$$

where φ is any stratified formula not containing x.

Then for example we can form the *set* $V = \{x \mid x = x\}$ and we shall have $V \in V$, but we cannot form the collection $\{x \mid x \notin x\}$ since the formula $x \notin x$ is not stratified. So the subset axiom must fail, as must replacement and the axiom of foundation. But we can form the set of all singletons, and the set of all doubletons, etc., and use those as our numbers one, two, etc., as Frege wanted to [Fre84]; in fact we can treat any cardinal number as the set of all sets equivalent to some given set.

Quine used this as a basis for reconstructing substantial parts of mathematics, but development of mathematics on this basis effectively ceased when it was shown that the axiom of choice could be disproved [Spe53]. The main interest in it is now the simple question of its consistency. It was shown by Jensen [Jen69] that if the axiom of extensionality is weakened to allow urelements other than the empty set (as in 3.2.3(1)), then the system is consistent; and there are further results, [BC85]. But it is still open whether the full system is consistent.

9.5.7 Exercise

(1) Show that the definition of *stratified* cannot be relaxed to allow labellings in which the variables are just ranked (with $x \in y$ allowed if the rank of x is less than that of y, rather than insisting that it be exactly one less).

[Show that the paradox of grounded sets will arise if the labels are allowed to be two apart, since the formula

$$\forall a(x \in a \Rightarrow \exists y \in a(y \cap a = \emptyset))$$

would then be allowed in an abstraction term.]

9.5.8 Mathematics without the axiom of foundation

We have pointed out, in various places, that the axiom of foundation, although a natural consequence of the concept of the cumulative types, is not needed in much of mathematics. Thus 6.5.8(2) shows that we have a satisfactory definition of ordinals (and hence of natural numbers, and so of integers, rationals, reals, etc.) without the axiom. We sketch here a proof, based on transfinite induction using these ordinals, of the relative consistency of the axiom of foundation. This proof was given by von Neumann [vN28b], using transfinite induction.

We let Ord x be Ord$''$ x as in 6.5.8(2) and first note that transfinite induction will then be available as in 6.6 even without foundation, and so we can repeat the definition of ranks from 6.6.13(3). So let V_α be given; and for convenience let V be introduced as a proper class $V = \bigcup_{\text{Ord}\,\alpha} V_\alpha$ with the convention that $x \in V$ abbreviates $\exists \alpha(x \in V_\alpha)$. Then the proof consists of showing that $\langle V, \in \rangle$ is a model of all the axioms of ZF, including the axiom of foundation, provided we assume the other axioms in the meta-language.

9.5.9 The formal presentation

At a formal level, we need the notion of *relativized* terms and formulas, from 8.2.3.

The theorem is:

Theorem 9.5.10 *If φ is an axiom of ZF (including the axiom of foundation), then there is a proof in ZF minus the axiom of foundation of φ^V; and similarly* for ZFC.

Proof All the axioms can be written with no abstraction terms, and the proof consists in writing equivalent forms of the axioms in which as many as possible of the quantifiers are bounded. The remaining quantifiers must then be examined; in most cases these are existential quantifiers. Often the axiom in its original form says that something exists with the property (and since the property is expressed with bounded quantifiers, it is absolute). So the task is to show that the thing is in fact in V, and this is usually straightforward. This covers the axiom of extensionality (which is a bounded formula when written carefully), the axioms of null-set, pair-set, sum-set, subset, and choice.

It gives the axiom of foundation also when we write it in the form

$$\forall a(\exists x \in a(x = x) \Rightarrow \exists x \in a \forall y \in a(y \notin x))$$

since we now need to show this for all $a \in V$, and if $a \in V$ then $a \in V_\alpha$ for some α. So let $A = \{\beta \mid a \cap V_\beta \neq \emptyset\}$ and note that if $a \neq \emptyset$ then A is a non-empty set of ordinals below α. So A has a least member α_0 (since the ordinals are, by definition now, well-ordered), and $a \cap V_{\alpha_0} \neq \emptyset$, and if $x \in a \cap V_{\alpha_0}$ then $x \cap a = \emptyset$. This says that the formula required holds in fact (that is, in the universe of all sets). But the formula is bounded, so its relativization to V holds also.

The axioms of power-set and replacement need a little consideration. We can write the power-set axiom as

$$\exists x [\forall y \in x \forall z \in y(z \in a) \land \forall y(\forall z \in y(z \in a) \Rightarrow y \in x)]$$

and of course we take x as $\mathcal{P}(a)$; if $a \in V_\alpha$ then $\mathcal{P}(a) \in V_{\alpha+1}$ and so is in V. But the second conjunct has an unbounded universal quantifier, and we must use the observation that if $a \in V$ and $y \subseteq a$ then $y \in V$ (this has been called the *supertransitive* property of V). This means that the second conjunct is also absolute, and so the relativization of the power-set axiom to V is proved.

For the replacement axiom, suppose that we wish to prove the instance of the replacement axiom for the formula $\psi(z, y)$, relativized to V. Then we start by using the instance of the replacement axiom for the formula $\psi^V(z, y)$, and we may as well assume that this is functional so that we have to show

$$\exists x \in V \forall y \in V(y \in x \Leftrightarrow \exists z \in a \psi^V(z, y)) \qquad (*)$$

given that

$$\exists x \forall y (y \in x \Leftrightarrow \exists z \in a \psi^V(z, y)).$$

If we let $X = \{y \mid \exists z \in a \psi^V(z, y)\}$ then $X \cap V = \{y \in V \mid \exists z \in a \psi^V(z, y)\}$ and is the set required to satisfy $(*)$. So we need to show $X \cap V \in V$.

MISCELLANEOUS FURTHER TOPICS   207

For $y \in V$ let $\rho(y)$ be the least α such that $y \in V_\alpha$ (in other words the rank of y), and let $A = \{\rho(y) \mid y \in X \cap V\}$. Then A is a set of ordinals, and if $\beta = \bigcup A$ then $X \cap V \in V_{\beta+1}$. □

9.5.11   Ackermann's set theory

W. Ackermann [Ack56] gave a system for set theory which has aspects of the class theories in that there is a class constant V and which allows arbitrary class construction for members of V (i.e. $\{x \in V \mid \varphi(x)\}$ is a class for any formula $\varphi(x)$). V is taken to be *supertransitive*, that is, not only members of members, but also subsets of members of V are members of V. Extensionality is assumed for all classes. The distinctive axiom is then *Ackermann's schema*: if $\varphi(x)$ is a formula which does not use the class constant V, and with free variables $x, y_1, \ldots, y_n$, then the following is an axiom:

$$\forall y_1, \ldots y_n \in V(\forall x(\varphi(x) \Rightarrow x \in V) \Rightarrow \exists w \in V \forall x(x \in w \Leftrightarrow \varphi(x)))$$

where w is not in φ. In other words, the class $\{x \in V \mid \varphi(x)\}$ is a member of V (and so a *set* in Ackermann's system).

It is a simple exercise, on the basis of these axioms, to show that the class V satisfies the set-existence axioms of Zermelo (that is, null-set, pair-set, sum-set, power-set and subset). But the status of the replacement axiom is not so clear, and was an open question for some years.

Eventually it was shown that if the axiom of foundation is added and we call the resulting system A^*, then A^* is equivalent to ZF, in a very strong sense. Suppose that φ is a formula of LST (so not involving V), and let φ^V be the result of relativizing all quantifiers to V. Then

$$A^* \vdash \varphi^V \quad \text{if and only if} \quad ZF \vdash \varphi.$$

These results are due to Levy [Lev59] and Reinhardt [Rei70].

9.5.12   Exercise

(1) Show that in Ackermann's system, the axioms of null-set, pair-set, sum-set, power-set and subset all hold for V, in the sense that if the starting sets are in V then so is the resulting set.

9.6   Remarks on the philosophy of mathematics

At various places in this book we have made remarks about the implications of Gödel's theorems for the work we are presenting; these concern such matters as the incompleteness of the axioms, and the lack of uniqueness of structures such as the natural numbers and the real numbers. We shall try to summarize here the basic implications and their import for the philosophy of mathematics.

Much (though by no means all) of the work described in this book was originally developed because of an interest in the foundations of mathematics and the philosophy of mathematics; the paradoxes described in 1.3 brought this interest to the forefront. To many, set theory was of interest because it was able to give answers to such questions as "what is a number?", and, when set theory had been formalized as an axiomatic theory, such questions as "what can and what cannot be proved about natural numbers?".

It is important to realize that this is not the only motivation for studying set theory, and it is far from the most important motivation today. It was not Cantor's motivation (he wanted to understand questions about the convergence of Fourier series, and had to study sets of reals to answer those questions). Much current work in set theory is concerned with independence: almost all branches of pure mathematics have led to questions which are independent of ZFC, and progress may be made by studying the question in specific forcing models (which is a convenient way of adding new properties). A separate, but related contribution to some of these questions is the development of set-theoretical methods which are often called *combinatorial* (or just technical), and which turn out to be essential extensions of more traditional methods. Transfinite induction is probably the first such method; of more modern developments, we have only been able to sketch some of the work on *stationary sets* in 7.4.6(8). More will be found e.g. in [She94].

However, this shift of interest from the historical concern with foundations does not mean that set theory is irrelevant to modern work on the foundations of mathematics; it would be impossible to contribute to the philosophy of mathematics without taking account of the position of modern set theory. It is the purpose of this section to set out that position.

9.6.1 *The completeness of first-order logic*

First, as noted in 2.4 when introducing first-order logic, is the certainty of formal, first-order proofs. There is no doubt about proofs from the axioms of ZFC; while there is often doubt about claimed proofs in mathematics, there is one clear way to resolve them which is universally accepted (if it can be carried out), namely to reduce to constructions and proofs within ZFC. It seems fair to claim that this is established to the point that many mathematicians will accept proofs within ZFC as the only established mathematics (even though they may want to use ideas which go beyond established methods).

This seems to apply to many who would regard set theory as a constraint on the methods to be used—for example, who would definitely want to study structure, as in category theory, and not specific constructions. They would still wish to use only methods which have been shown to be reducible to ZFC.

MISCELLANEOUS FURTHER TOPICS

But others definitely do want to get away from such set-theoretic constraints (as they see them). This brings us to the second point.

9.6.2 Gödel's incompleteness theorems

The first theorem can be stated as: no formal theory in which arithmetic can be expressed can be both consistent and complete. (How much of arithmetic is needed, need not concern us here since even the weakest set theories considered are strong enough for the theorem to apply.)

The second theorem will also apply to all the set theories we have considered, and this says: provided the consistency of a theory is formalized suitably, then this consistency will be provable within the theory only if that theory is in fact inconsistent. (Remember that within classical theories, a theory is inconsistent if and only if it can prove *every* statement of the theory.)

This second Gödel theorem means that even if other methods can be developed in most powerful and attractive ways, the question will still be asked, what can be said about the consistency of the methods being used? It is hard to imagine methods being attractive to large numbers of mathematicians if they are too weak to allow Gödel's theorem to apply. So the only hope for progress on the question of consistency is to compare consistency strength with that of other more established theories, and this in practice means set theory as generally accepted—hence with ZFC, or its strengthenings by large cardinals.

An extreme illustration of this last aspect of set theory is the attitude taken toward the set theories of Quine and of Ackermann (as in 9.5.5 and 9.5.11). Very little work has been done in Ackermann's set theory, since Reinhardt settled the question of its consistency strength by showing it equivalent to ZFC; and almost the only current interest in Quine's set theory is in the question of its consistency.

9.6.3 Consequences of the first incompleteness theorem

Much the most obvious aspect of set theory to the rest of mathematics today is this incompleteness: as noted above, many questions of interest to other mathematicians have turned out to be unanswerable on the basis of ZFC, because ZFC is incomplete. Since Cohen [Coh63a] many different forcing models have been given, from which these independence results follow.

The nature of Gödel's proof shows that this is inescapable. We might hope to find further intuitions into the heuristic structure, which could be made into further axioms. Work on large cardinals and on the axiom of determinacy may well have been inspired by such aims. We could hope for completely different intuitions on which to base mathematics. But so long as we remain with a formal system—and this will be essential if others are to be able to check our proofs—we shall have an incomplete system (or an inconsistent one), and independence will never go away. The current situation, in which all possible combinations of forcing models and large cardinals are considered, may well be refined by agreement about which are reasonable and which are bizarre. This would be an example of further intuition; it seems more likely to come from investigating many diverse possibilities than from anywhere else.

A further consequence of the first incompleteness theorem, is the existence of non-standard models. This has been exploited in various ways, particularly in analysis and probability theory, following Abraham Robinson (e.g. [Rob66]). Essentially this is always the exploitation of the fact that the first-order theories of interest are strong enough to admit Gödel's first incompleteness theorem, and so have non-isomorphic models. These must be different from the intended, standard model, and yet still have the properties which are provable from the axioms; this gives new ways to argue and to think about the theories. The very existence of the subject is a reminder to consider what we are doing when we repeat, for example, Dedekind's proof of the uniqueness of the natural numbers, as in 6.2.11.

9.6.4 Summary of the realist or platonist position

This is perhaps simplest stated as: there is a unique structure, the cumulative type structure of sets. This structure contains standard copies of the integers, the real numbers, etc. Mathematics is the study of these standard structures, and ZFC is just the best agreed first-order approximation to this standard structure, giving the best information we have about the integers, real numbers, etc.

Gödel's theorems then place severe limitations on our ability to complete this study of mathematics. However good a first-order approximation we achieve, it will not be complete and will have non-standard models and independence results.

9.6.5 Other standpoints

Traditionally the other standpoints on the foundations of mathematics were classified as *formalist* or as *constructivist*.

The constructivist standpoint could be stated as: mathematics is the study of constructions. Some properties of these constructions may lead to agreements about formal statements, but not necessarily within classical first-order logic (usually the logic will reject the *excluded middle*; propositions are no longer necessarily either true or false). The original rejection of formalization by Brouwer has not been followed by later workers, and it is possible to draw a parallel between the justification of axioms for sets by platonists from the heuristic picture of the cumulative types, and the deduction of formal axiom systems from the heuristic notion of a construction by various schools of constructivists. It is then possible to compare the consistency strength of the resulting systems: in general the constructively justified systems will be very much weaker than ZFC.

The formalist standpoint is generally associated with Hilbert, and an extreme form would restrict mathematics to the study of formal systems alone, rejecting any attempt at justification of them. (It seems certain that this was not Hilbert's view.) It is sometimes caricatured as the statement that mathematics is the same as playing games (formal axiom systems being simply the rules for a game, rather than having any external meaning). Hilbert's programme had as its aim the proof of the consistency of formal systems such as ZFC; but this was shown impossible by Gödel, and it is not clear whether there is any agreed formalist standpoint today. Certainly there are mathematicians who reject anything as being mathematics unless it consists of derivations which can be formalized in some formal system. For them, questions of justification of formal systems lie outside of mathematics. But it is hard to see this as other than a restrictive definition, which places the justification of formal systems perhaps as applied mathematics, or philosophy, or aesthetics; after all, some choice of formal system has to be made before any mathematics can happen, on the formalist view.

There are many variants of the positions sketched above, and there are other positions such as strict finitism (which will deny any reality to infinite objects, but which may be a realist position with regard to at least certain finite sets), and nominalism, which insists that language is primary and all that can be studied. But this is only intended as a brief introduction, to encourage the reader to consider these questions; further references are [BP64], [vH67].

10

Appendix: Some basic definitions

This listing is intended for reference; there are many variants in the way that these definitions can be given, and we give some of those variants. The main aim is to enable the reader to follow the development in the main text, and this is clearly hampered if we use a notation which is not familiar to the reader. However we do assume that most of these will have been met before.

10.1 Simplest constructions, and variants

10.1.1 Abstraction terms

(i) $\{x \mid \varphi\}$ for the abstraction term to represent the collection (class) of all x's which have property φ (which may or may not be a set).

Alternatives: $\{x\ ;\ \varphi\}$ or $\{x : \varphi\}$. (These are possibly more common than $\{x \mid \varphi\}$. We prefer the latter since the vertical line | is more visible, particularly on a blackboard, than the semi-colon or colon.) Less common: $\widehat{x}\varphi$.

(ii) $\emptyset$ for the empty set, which has no members. Alternative: Λ (not very common in set theory, more often used in lattice theory, or to denote the empty sequence, which is the same thing as the empty set to a set-theorist).

(iii) $\{t \mid \varphi\}$ (where t is an abstraction term itself, and not just a variable), for the collection of t's such that φ, i.e. the class (which may or may not be a set):

$$\{x \mid \exists x_1, \ldots, x_n (x = t(x_1, \ldots, x_n) \wedge \varphi(x_1, \ldots, x_n))\},$$

where $x_1, \ldots, x_n$ lists all the variables common to t and φ. See 10.2.1(iii) and 10.2.3(iv) below for examples of this usage, and also the usage for functions which corresponds in 10.2.3(vi).

There may be also variants in which not all the common variables are to be quantified, but some are left as parameters, that is, free variables which the term $\{t \mid \varphi\}$ then depends on. We have tried to avoid this, but it is sometimes a natural construction.

10.1.2 Boolean operations, etc.

(i) $a \cup b$ for the union of a and b, i.e. $\{x \mid x \in a \vee x \in b\}$.

(ii) $a \cap b$ for the intersection of a and b, i.e. $\{x \mid x \in a \wedge x \in b\}$.

(iii) $a - b$ for the complement of b within a, i.e. $\{x \mid x \in a \wedge x \notin b\}$. Alternative: $a \setminus b$.

(iv) $a \Delta b$ for the symmetric difference of a and b, i.e. $\{x \mid (x \in a \wedge x \notin b) \vee (x \in b \wedge x \notin a)\}$, i.e. $(a - b) \cup (b - a)$. Alternatives: various; e.g. $a \oplus b$ and $a + b$ have been used. Also Δ has been used for diagonal intersections, see 7.4.6(8).

(v) $a \subseteq b$ for: a is a subset of b, i.e. $\forall x (x \in a \to x \in b)$. Alternative: $a \subset b$. At some stage there must have been reluctance to admit the necessity for distinguishing the strict subset relation (proper subset), from the non-strict relation (which allows equality); and the symbol $\subset$ was and still is widely used in the non-strict sense. To be safe, most authors feel the need to write something like $\subsetneq$ or $\subsetneqq$ for the strict relation.

Note that the English word *contains* is a source of confusion for beginners in set theory, since it can mean either contains *as a subset* or contains *as a member*. But *strictly contains* is not ambiguous; that has to mean contains *as a proper subset*.

(vi) $\mathcal{P}(a)$ for the power-set of a, i.e. $\{x \mid x \subseteq a\}$. Alternative: $\mathbb{P}(a)$.

(vii) $\bigcup a$ for the union (or sum-set) of the set a, i.e. the union of all the members of a, which is $\{x \mid \exists y (x \in y \wedge y \in a)\}$.

(viii) $\bigcap a$ for the intersection of the set a, i.e. the intersection of all the members of a, which is $\{x \mid \forall y (y \in a \to x \in y)\}$. Warning: note this is a proper class (the universal class, V) if a is empty.

Alternatives (older, and not common today): $\sum a$ for union, $\prod a$ for intersection. Also see under families, 10.2.4.

10.2 Ordered pairs, relations, functions, families and sequences

10.2.1 Ordered pairs

(i) $\langle a, b \rangle$ for the ordered pair of a and b, i.e. $\{\{a\}, \{a, b\}\}$ (where $\{a, b\}$ is the unordered pair $\{x \mid x = a \vee x = b\}$).

APPENDIX: SOME BASIC DEFINITIONS

The basic property of ordered pairs is in exercise 10.2.5(1). Note that it was not until the discovery of a definition of ordered pairs in terms of sets alone, that *set-theoretic reductionism* (the claim that all of mathematics is set theory) could really be considered. The next definitions were also essential steps towards the reduction of mathematics to set theory, but these steps were taken before the ordered pair was reduced as above.

(ii) $\langle a, b, c \rangle$ for the ordered triple, i.e. $\langle \langle a,b \rangle, c \rangle$; and similarly for ordered quadruples, quintuples, etc. Alternative: $\langle a, \langle b, c \rangle \rangle$. See the comments on domain and range below.

(iii) $a \times b$ for the cartesian product of a and b, i.e. $\{\langle x, y \rangle \mid x \in a \wedge y \in b\}$. Similarly for $a \times b \times c$, etc.

10.2.2 Relations

(i) Rel **r** for: **r** is a relation, i.e. a set of ordered pairs. So

$$\text{Rel}\, \mathbf{r} \Leftrightarrow \forall x (x \in \mathbf{r} \to \exists u, v (x = \langle u, v \rangle)).$$

(ii) dom **r** for the domain of **r**, i.e. $\{x \mid \exists y (\langle x, y \rangle \in \mathbf{r})\}$.

(iii) ran **r** for the range of **r**, i.e. $\{x \mid \exists y (\langle y, x \rangle \in \mathbf{r})\}$.

Alternative: do be aware that there is roughly 50% chance that an author will switch these two. The same applies to the definition of ordered triples, etc. The choice corresponds to whether the argument comes before or after the value, in a function. Usually it will not be very important, but simply a matter of switching the order in some definition or statement, if things get mixed; so most mathematicians are understandably careless in this respect. Of course, it usually *does* matter which we are referring to, the argument or the value of a function; and there have been a few occasions when it was left unclear which was intended, because an author relied on formal statements without explanation, and typing errors in the formal statements rendered them unintelligible. Explanations can be far more valuable than adjustments to formal notation.

(iv) $u\, \mathbf{r}\, v$ for: relation **r** holds between u and v, i.e. $\langle u, v \rangle \in \mathbf{r}$.

(v) $\mathbf{r}^{-1}$ for the converse relation, i.e. $\{\langle u, v \rangle \mid \langle v, u \rangle \in \mathbf{r}\}$ or $\{\langle u, v \rangle \mid v\, \mathbf{r}\, u\}$. Alternative: $\mathbf{r}^{\cup}$.

(vi) $\mathbf{r} \upharpoonright z$ for the restriction of relation **r** to the domain z (or really to the domain $(\text{dom}\, \mathbf{r}) \cap z$, i.e. $\{\langle u, v \rangle \mid u\, \mathbf{r}\, v \wedge u \in z\}$ or $\mathbf{r} \cap (z \times \text{ran}\, \mathbf{r})$.

(vii) $\mathbf{r} \circ \mathbf{s}$ for the relational product of **r** and **s**, i.e. $\{\langle x, z \rangle \mid \exists y (x\, \mathbf{r}\, y \wedge y\, \mathbf{s}\, z)\}$.

10.2.3 Functions

Note that to a set-theorist, an ordered pair is a set, just as much as any other set, and functions of more than one variable are simply functions whose arguments happen to be ordered pairs, or triples, etc., so no special definitions are needed to handle them. All the definitions will deal with functions of one variable, ostensibly, but will apply to functions of any number of variables. The same should be noted about relations: ternary (three-place) relations can be regarded as binary relations with ordered pairs in the domain (or range), etc.

(i) Fn f for: f is a function, i.e. a relation which is *functional*; that is, to each argument of the domain there is just one value. So

$$\text{Fn } f \Leftrightarrow \text{Rel } f \wedge \forall x \in \text{dom } f \exists^1 y (\langle x, y \rangle \in f).$$

Alternative: Func f.

(ii) $f : x \to y$ for: f is a function from x to y, i.e. Fn $f \wedge \text{dom } f = x \wedge \text{ran } f \subseteq y$.

(iii) $y = f(x)$ for: f is a function, $x \in \text{dom } f$, and $\langle x, y \rangle \in f$. Alternatives: $f : x \mapsto y$ or $f : x \rightsquigarrow y$.

Note that some authors may relax the demand that f be a function, and demand only that f be functional for x. Then the definition can be taken as

$$f(x) = \{z \mid \exists y (\forall w (\langle x, w \rangle \in f \Leftrightarrow w = y) \wedge z \in y)\}.$$

But note that this will give $f(x) = \emptyset$ for any x not in dom f or for which f is not functional; it is not clear that this is always wanted. We shall often include the condition Fn f in our definitions so that these spurious values cannot interfere. Another alternative notation used by Russell is $y = f'x$; for yet another, more common, see under *Families and sequences*.

(iv) $f''z$ for the image of z under f, i.e. $\{f(x) \mid x \in z\}$, the set of values of f for arguments in z. Note that $f''z = \text{ran } f \restriction z$. Alternatives: there are many, e.g. $f * z$, $f[\![z]\!]$, and even $f(Z)$, in cases where it is clear that Z could not be an argument for f but must be treated as a set of arguments (in which case the arguments for f would be denoted by lower-case letters). We must treat this usage with care within set theory, where *every* object is a set; but it is very common, and sometimes natural, as in the proof of 4.2.6.

(v) $^Y X$ for the set of functions from Y to X, i.e. $\{f \mid f : Y \to X\}$.

Alternative: X^Y is sometimes used for this.

(vi) $\langle t \mid \varphi \rangle$ for the function taking value t for each argument satisfying φ, i.e. $\{\langle x, t \rangle \mid \varphi(x)\}$ where x is the free variable common to both t and φ. If there are two or more such free variables, then this would denote a function of two or more variables, and in $\langle x, t \rangle$ the x would be an ordered pair, or triple, etc.

(vii) $X \sim Y$ for: X and Y are in 1–1 correspondence (are *equivalent*, as in 4.1.1). Alternatives: $X \approx Y$ or $X \simeq Y$.

APPENDIX: SOME BASIC DEFINITIONS

10.2.4 Families and sequences

Both *families* and *sequences* are taken as particular sorts of functions. A *family* could be any function, but the domain will then be called the *index-set* of the family and the values will be called the *members* of the family, *indexed* by the appropriate argument. A *sequence* is simply a family or function whose domain happens to be an ordinal, and the members of the sequence are then the values of the function. An *infinite sequence* is usually one with domain ω.

(i) X_i for the i-th member of family X, or the i-th member of sequence X, i.e. $X(i)$.

(ii) $(X_i)_{i \in I}$ for the family (or sequence if I is an ordinal), with index-set I and member X_i at the index $i \in I$, i.e. the function $\langle X(i) \mid i \in I \rangle$.

(iii) $\bigcup_{i \in I} X_i$ for the union of the family $(X_i)_{i \in I}$, i.e. $\bigcup \{X_i \mid i \in I\}$. Similarly for $\bigcap_{i \in I} X_i$.

(iv) $\bigtimes_{i \in I} X_i$ for the cartesian product of the family $(X_i)_{i \in I}$, i.e.

$$\{f \mid f : I \to \bigcup_{i \in I} X_i \wedge \forall i \in I (f(i) \in X_i)\}.$$

Alternative: $\prod_{i \in I} X_i$ (but note that $\sum_{i \in I} X_i$ was at one time used for $\bigcup_{i \in I} X_i$, and $\prod_{i \in I} X_i$ for $\bigcap_{i \in I} X_i$; this confusion is a good reason to leave the symbols $\sum$ and $\prod$ to mean only addition and multiplication.

10.2.5 Exercises

(1) Prove the basic property of ordered pairs, i.e. $\langle a, b \rangle = \langle c, d \rangle \Leftrightarrow a = c \wedge b = d$, taking care to include all possible cases.

Show that an alternative definition of ordered pair is given by $(a, b) = \{\{a, 0\}, \{b, 1\}\}$ provided $0 \neq 1$. But show that this cannot be continued to give an ordered triple as $(a, b, c) = \{\{a, 0\}, \{b, 1\}, \{c, 2\}\}$.

[Show that $(1, 2, 0)$ would be ambiguous.]

(2) Show that all the terms constructed in 10.1 and 10.2 give *sets* (and not proper classes), with the exception of the general abstractions $\{x \mid \varphi\}$, $\{t \mid \varphi\}$, and $\langle t \mid \varphi \rangle$ (which can, of course, give proper classes if φ is too inclusive); and also $\bigcap \emptyset$ as noted in 10.1.2(viii).

[These proofs must all start from the assumption that the free variables involved all denote *sets*, and this is of course assumed in the way the axioms are presented in chapter 3.]

(3) Show that of these terms which can be proved to be sets, only the power-set itself 10.1.1(vi), the function set $^Y X$ in 10.2.3(v), and the cartesian product of a family $\bigtimes_{i \in I} X_i$ in 10.2.4(iv), make essential use of the power-set axiom.

[For example, to get the simple cartesian product $a \times b$, one proof would be to note that it is a subset of $\mathcal{P}(\mathcal{P}(a \cup b))$. But this is making strong use of the power-set axiom, and in some contexts it is important to know that this can be done instead using the replacement axiom. Show first, using replacement on b, that $\{i\} \times b$ is a set for each $i \in a$. Then use replacement on a to show that $\{\, \{i\} \times b \mid i \in a\,\}$ is a set, and then use the sum-set axiom.

To show that the power-set axiom is in fact needed, for the three constructions noted, the model of the hereditarily countable sets can be used as in 6.7.8(3).]

(4) Show that the direct definition of function value given in 10.2.3(iii) cannot be written as
$$f(x) = \{z \mid \exists^1 y(\langle x, y \rangle \in f \wedge z \in y)\}.$$

[But it could be written
$$f(x) = \{z \mid \exists^1 y(\langle x, y \rangle \in f) \wedge \exists^1 y(\langle x, y \rangle \in f \wedge z \in y)\}.]$$

References

[Ack56] Ackermann W. (1956) Zur Axiomatik der Mengenlehre. *Math Ann* 131: 336–345.
[AHT65] Addison J. W., Henkin L., and Tarski A. (eds) (1965) *The theory of models. Proceedings of the 1963 international symposium at Berkeley.* Studies in Logic and the Foundations of Mathematics. North-Holland.
[Bac55] Bachman H. (1955) *Transfinite Zahlen.* Springer Verlag.
[Bar77] Barwise J. (ed) (1977) *Handbook of mathematical logic.* North-Holland. Second edition 1978.
[BC85] Boffa M. and Casalegno P. (1985) The consistency of some 4-stratified subsystems of NF including NF_3. *Journal of symbolic logic* 50: 407–411.
[Bel85] Bell J. L. (1985) *Boolean-valued models and independence proofs in set theory*, volume 12 (Second edition) of *Oxford Logic Guides.* Oxford University Press.
[Ber54] Bernays P. (1937–1954) A system of axiomatic set theory, parts i–vii. *Journal of symbolic logic* 2–19. Reprinted in [Mül76].
[BF97] Burali-Forti C. (1897) Una questione sui numeri transfiniti. *Rend Circ Mat Palermo* 11: 154–164. Reprinted in [vH67].
[BJW82] Beller A., Jensen R. B., and Welch P. (1982) *Coding the universe*, volume 47 of *London Mathematical Society Lecture notes.* Cambridge University Press.
[Bla87] Blass A. R. (ed) (1987) Ω-*Bibliography of mathematical logic, volume 5, Set theory.* See [Mül87].
[BP64] Benacerraf P. and Putnam H. (eds) (1964) *Philosophy of mathematics, selected readings.* Prentice-Hall philosophy series. Prentice-Hall.
[Can32] Cantor G. (1932) *Gesammelte Abhandlungen.* Springer Verlag. Second edition 1980.
[CK73] Chang C.-C. and Keisler H. J. (1973) *Model theory.* Studies in Logic and the Foundations of Mathematics. North-Holland. Third edition 1990.
[Coh63a] Cohen P. J. (1963) The independence of the continuum hypothesis. *Proceedings of the National Academy of Sciences USA* 50: 1143–1148.
[Coh63b] Cohen P. J. (1963) Independence results in set theory. In Addison *et al.* [AHT65], pages 39–54.
[Coh64] Cohen P. J. (1964) The independence of the continuum hypothesis ii. *Proceedings of the National Academy of Sciences USA* 51: 105–110.
[Coh66] Cohen P. J. (1966) *Set theory and the continuum hypothesis.* W. A. Benjamin.
[Dav64] Davis M. (1964) Infinite games of perfect information. *Annals of Mathematics Studies* 52: 85–101.

[Dav65] Davis M. (ed) (1965) *The undecidable; basic papers on undecidable propositions, unsolvable problems and computable functions.* Raven Press.
[Ded48] Dedekind R. (1948) *Essays on the theory of numbers.* Dover.
[Ded88] Dedekind R. (1888) *Was sind und was sollen die Zahlen?* Vieweg. Translated in [Ded48].
[Dev84] Devlin K. J. (1984) *Constructibility.* Perspectives in Mathematical Logic. Springer-Verlag.
[DJ81] Dodd A. J. and Jensen R. B. (1981) The core model. *Annals of Mathematical Logic* 20: 43–75.
[Eas70] Easton W. (1970) Powers of regular cardinals. *Annals of Mathematical Logic* 1: 139–178.
[EHMR84] Erdős P., Hajnal A., Máté A., and Rado R. (1984) *Combinatorial set theory.* Studies in Logic and the Foundations of Mathematics. North-Holland.
[End72] Enderton H. B. (1972) *A mathematical introduction to logic.* Academic Press.
[Fod56] Fodor G. (1956) Eine Bemerkung zur Theorie der regressiven Funktionen. *Acta Sci Math (Szeged)* 17: 139–142.
[Fra22a] Fraenkel A. A. (1922) Axiomatische Begrundung der transfinite Kardinalzahlen I. *Math Z* 13: 153–188.
[Fra22b] Fraenkel A. A. (1922) Zu den Grundlagen der Cantor-Zermeloschen Mengenlehre. *Math Ann* 86: 230–237.
[Fre03] Frege F. L. G. (1903) *Grungezetze der Arithmetik, Begriffsschriftlich abgeleitet, vol2.* Pohle.
[Fre84] Frege F. L. G. (1884) *Die Grundlagen der Arithmetik, eine logisch mathematische Untersuchung über den Begriff der Zahl.* Köbner.
[Fre93] Frege F. L. G. (1893) *Grungezetze der Arithmetik, Begriffsschriftlich abgeleitet, vol1.* Pohle.
[Frem84] Fremlin D. H. (1984) *Consequences of Martin's axiom.* Cambridge University Press.
[Göd38] Gödel K. (1938) The consistency of the axiom of choice and of the generalized continuum-hypothesis. *Proceedings of the National Academy of Sciences USA* 24: 556–557.
[Göd39] Gödel K. (1939) Consistency-proof for the generalized continuum-hypothesis. *Proceedings of the National Academy of Sciences USA* 25: 220–224.
[Göd40] Gödel K. (1940) *The consistency of the axiom of choice and of the generalized continuum-hypothesis with the axioms of set theory*, volume 3 of *Annals of Mathematics Studies*. Princeton University Press.
[Hal84] Hallett M. (1984) *Cantorian set theory and the limitation of size*, volume 10 of *Oxford Logic Guides*. Clarendon Press.
[Ham88] Hamilton A. G. (1988) *Logic for mathematicians.* Cambridge University Press, revised edition.
[Har15] Hartogs F. (1915) Über das Problem der Wohlordnung. *Math Ann* 76: 438–443.
[Hau35] Hausdorff F. (1935) *Mengenlehre.* Gruyter. Translated 1975. Chelsea.
[HMSS85] Harrington L. A., Morley M., Scedrov A., and Simpson S. G. (eds) (1985) *Harvey Friedman's research on the foundations of mathematics.* North-Holland.
[HP77] Harrington L. A. and Paris J. B. (1977) A mathematical incompleteness in Peano Arithmetic. In Barwise [Bar77]. Second edition 1978.
[Jec73] Jech T. J. (1973) *The axiom of choice.* Studies in Logic and the Foundations of Mathematics. North-Holland.
[Jec74] Jech T. J. (ed) (1974) *Axiomatic set theory. Proceedings of the 1967*

symposium at UCLA. Part 2, volume 13 of *Proceedings of Symposia in Pure Mathematics*. American Mathematical Society.

[Jen69] Jensen R. B. (1969) On the consistency of a slight (?) modification of Quine's New Foundations. *Synthese* 19: 250–263.

[Jen72] Jensen R. B. (1972) The fine structure of the constructible hierarchy. *Annals of Mathematical Logic* 4: 229–308.

[Kan94] Kanamori A. (1994) *The higher infinite*. Perspectives in Mathematical Logic. Springer-Verlag.

[Kec95] Kechris A. S. (1995) *Classical descriptive set theory*. Graduate Texts in Mathematics. Springer-Verlag.

[Kel55] Kelley J. L. (1955) *General topology*. Van Nostrand. Later edition Springer Verlag.

[Kön05] König J. (1905) Über die Grundlagen der Mengenlehre und das Kontinuumsproblem. *Math Ann* 61: 156–160. Reprinted in [vH67].

[Kön27] König D. (1927) Über eine Schlussweise aus dem Endlichen ins Unendliche. *Acta Univ Szeged, Sect Mat* 3: 121–130.

[Kun80] Kunen K. (1980) *Set theory: an introduction to independence proofs*. Studies in Logic and the Foundations of Mathematics. North-Holland.

[KV84] Kunen K. and Vaughan J. E. (eds) (1984) *Handbook of set-theoretic topology*. North-Holland.

[Lev59] Levy A. (1959) On Ackermann's set theory. *Journal of symbolic logic* 24: 154–166.

[Lev60] Levy A. (1960) Axiom schemata of strong infinity in axiomatic set theory. *Pacific Journal of Mathematics* 10: 223–238.

[Lev65a] Levy A. (1965) The Fraenkel-Mostowski method for independence proofs in set theory. In Addison et al. [AHT65], pages 221–228.

[Lev65b] Levy A. (1965) *A hierarchy of formulas in set theory*, volume 57 of Memoirs of the American Mathematical Society. American Mathematical Society. Reprinted 1974.

[Lev79] Levy A. (1979) *Basic set theory*. Perspectives in Mathematical Logic. Springer-Verlag.

[LS67] Levy A. and Solovay R. M. (1967) Measurable cardinals and the continuum hypothesis. *Israel Journal of Mathematics* 5: 234–248.

[Mat83] Mathias A. R. D. (ed) (1983) *Surveys in set theory*, volume 87 of London Mathematical Society Lecture Notes. Cambridge University Press.

[Men64] Mendelson E. (1964) *Introduction to mathematical logic*. Van Nostrand Reinhold. Re-published by Springer Verlag.

[Mir17] Mirimanoff D. (1917) Remarques sur la théorie des ensembles et les antinomies Cantoriennes i. *Enseign Math* 19: 209–217.

[Mir20] Mirimanoff D. (1920) Remarques sur la théorie des ensembles et les antinomies Cantoriennes ii. *Enseign Math* 21: 29–52.

[Mon55] Montague R. M. (1955) On the paradox of grounded classes. *Journal of symbolic logic* 20: 140.

[Moo82] Moore G. H. (1982) *Zermelo's axiom of choice. Its origin, development and influence*. Springer Verlag.

[Mor65] Morse A. P. (1965) *A theory of sets*. Academic Press.

[Mos80] Moschovakis Y. N. (1980) *Descriptive set theory*. Studies in Logic and the Foundations of Mathematics. North-Holland.

[Mos94] Moschovakis Y. N. (1994) *Notes on set theory*. Undergraduate texts in mathematics. Springer-Verlag.

[MS62] Mycielski J. and Steinhaus H. (1962) A mathematical axiom contradicting

the axiom of choice. *Bull Acad Polon Sci.* 10: 1–3.
[MS70] Martin D. A. and Solovay R. M. (1970) Internal Cohen extensions. *Annals of Mathematical Logic* 2: 143–178.
[Mül76] Müller G. H. (ed) (1976) *Sets and classes. On the work of Paul Bernays.* Studies in Logic and the Foundations of Mathematics. North-Holland.
[Mül87] Müller G. H. (ed) (1987) Ω-*Bibliography of mathematical logic*, volume 1–5. Springer Verlag. edited in collaboration with W. Lenski.
[Pea89] Peano G. (1889) *The principles of arithemetic, presented by a new method (Latin)*. Bocca and Clausen. Reprinted in [vH67].
[Qui37] Quine W. v. O. (1937) New foundations for mathematical logic. *Amer Math Monthly* 44: 70–80.
[Qui63] Quine W. v. O. (1963) *Set theory and its logic*. Harvard University Press.
[Ram26] Ramsey F. P. (1926) The foundations of mathematics. *Proceedings of the London Mathematical Society series 2*, 25: 338–384.
[Rei70] Reinhardt W. N. (1970) Ackermann's set theory equals ZF. *Annals of Mathematical Logic* 2: 189–249.
[Rob66] Robinson A. (1966) *Non-standard analysis*. Studies in Logic and the Foundations of Mathematics. North-Holland. Second edition 1968.
[RS53] Rasiowa H. and Sikorski R. (1953) Algebraic treatment of the notion of satisfiability. *Fundamenta Mathematicae* 40: 62–95.
[Rus06] Russell B. R. (1906) On some difficulties in the theory of transfinite numbers and order types. *Proceedings of the London Mathematical Society series 2*, 4: 29–55.
[Sco57] Scott D. S. (1957) The notion of rank in set theory. In *Proceedings of the Summer Institute for Symbolic Logic, 1957, Ithaca*, pages 267–269. Institute for Defence Analyses, Communications Research Division.
[Sco67] Scott D. S. (1967) A proof of the independence of the continuum hypothesis. *Math Syst Theory* 1: 89–111.
[Sco71] Scott D. S. (ed) (1971) *Axiomatic set theory. Proceedings of the 1967 symposium at UCLA, part 1*, volume 13 of *Proceedings of Symposia in Pure Mathematics*. American Mathematical Society.
[Sco74] Scott D. S. (1974) Axiomatizing set theory. In Jech [Jec74], pages 207–214.
[She82] Shelah S. (1982) *Proper forcing*. Springer-Verlag.
[She94] Shelah S. (1994) *Cardinal arithmetic*, volume 29 of *Oxford Logic Guides*. Clarendon Press.
[Sho67] Shoenfield J. R. (1967) *Mathematical logic*. Addison-Wesley.
[Sho71] Shoenfield J. R. (1971) Unramified forcing. In Scott [Sco71], pages 357–381.
[Sie58] Sierpinski W. (1958) *Cardinal and ordinal numbers*, volume 34 of *Monograf. Mat.* PWN.
[Sko23] Skolem T. A. (1923) Einige Bemerkungen zur axiomatischen Begründung der Mengenlehre. In *Skand Mat Kongr (5) 1922 Helsinki*, pages 217–232. Reprinted in [vH67].
[Sko30] Skolem T. A. (1930) Einige Bemerkungen zu der Abhandlung von E. Zermelo: Über die Definitheit in Axiomatik. *Fundamenta Mathematicae* 15: 337–341.
[Sol65] Solovay R. M. (1965) $2^{\aleph_0}$ can be anything it ought to be. In Addison et al. [AHT65].
[Sol70] Solovay R. M. (1970) A model of set theory in which every set of reals is Lebesgue measurable. *Ann. of Math.* 92: 1–56.
[Spe53] Specker E. (1953) The axiom of choice in Quine's New Foundations for mathematical logic. *Proceedings of the National Academy of Sciences USA* 39:

972–975.

[ST71] Solovay R. M. and Tennenbaum S. (1971) Iterated Cohen extensions and Souslin's problem. *Ann Math Ser 2* 94: 201–245.

[Tar33] Tarski A. (1933) On the notion of truth in reference to formalized deductive sciences (Polish). *CR Soc Sci Lett Varsovie Cl 3* 34. Translated in [Tar56].

[Tar56] Tarski A. (1956) *Logic, semantics, metamathematics. Papers from 1923 to 1938 by Alfred Tarski.* Clarendon Press.

[Tei39] Teichmuller O. (1939) Braucht der Algebraiker das Auswahlaxiom? *Dt Math* 4: 567–577.

[vH67] van Heijenoort J. (ed) (1967) *From Frege to Gödel: a sourcebook in mathematical logic 1879–1931.* Harvard University Press. Second edition 1971.

[vN23] von Neumann J. (1923) Zur Einfuhrung der transfiniten Zahlen. *Act Univ Szeged, Sect Mat* 1: 199–208. Reprinted in [vH67].

[vN25] von Neumann J. (1925) Eine Axiomatisierung der Mengenlehre. *J Reine Angew. Math.* 154: 219–240. Reprinted in [vH67].

[vN28a] von Neumann J. (1928) Die Axiomatisierung de Mengenlehre. *Math Z* 27: 669–752.

[vN28b] von Neumann J. (1928) Über die Definition durch transfinite Induktion und verwandte Fragen der allgemeinen Mengenlehre. *Math Ann* 99: 373–391.

[WR10] Whitehead A. N. and Russell B. R. (1910) *Principia Mathematica.* Cambridge University Press. Volume 1 1910, Volume 2 1912, Volume 3 1913.

[YY29] Young W. H. and Young G. C. (1929) Review of E. W. Hobson, Second edition of The theory of functions of a real variable and the theory of Fourier's series. *Math Gazette* 14: 98–104.

[Zer04] Zermelo E. (1904) Beweis, dass jede Menge wohlgeordnet werden kann. *Math Ann* 59: 514–516. Reprinted in [vH67].

[Zer08] Zermelo E. (1908) Neuer Beweis für die Moglichkeit der Wohlordnung. *Math Ann* 65: 107–128. Reprinted in [vH67].

[Zer30] Zermelo E. (1930) Über Grenzzahlen und Mengenbereiche. Neue Untersuchungen über die Grundlagen der Mengenlehre. *Fundamenta Mathematicae* 16: 29–47.

[Zor35] Zorn M. A. (1935) A remark on a method in transfinite algebra. *Bulletin of the American Mathematical Society* 41: 667–670.

Index

$\in$, 15
$\Rightarrow$, 16
$\neg$, 16
$\wedge$, 16, 17, 56
$\vee$, 16, 17, 56
$\Leftrightarrow$, 16, 17
$\equiv$, 16
$\exists$, 16
$\forall$, 16, 17
$\exists^1$, 17
$:=$, 17
$\models$, 20
$\sim$, 37, 216
$\preccurlyeq$, 39
$\prec$, 39
$\aleph_0$, 47
$\aleph$, 47, 105
$\sharp$, 47
$|A|$, 47
$\overline{\overline{A}}$, 47
$\mathfrak{c}$, 47
$\simeq$, 53, 216
$\bigwedge$, 56
$\bigvee$, 56
$\prod'$, 68
$\dot{+}$, 72
ω, 73
$\restriction$, 98, 215
$'$, 98, 216
$<_L$, 145
x^G, 157
$\Vdash$, 158
$\check{\,}$, 159
$\Vdash^*$, 171
$[\ ]^{\mathbb{B}}$, 189
$\beth$, 200
$\{x \mid \varphi\}$, 213
$\emptyset$, 213
Λ, 213

$\hat{\,}$, 213
$\{t \mid \varphi\}$, 213
$\cup$, 214
$\cap$, 214
$-$, 214
$\setminus$, 214
$\triangle$, 214
$\oplus$, 214
$\subseteq$, 214
$\subset$, 214
$\subsetneq$, 214
$\bigcup$, 214, 217
$\bigcap$, 214, 217
$\sum$, 214, 217
$\prod$, 214, 217
$\langle\ ,\ \rangle$, 214
$\times$, 215
$u\,\mathbf{r}\,v$, 215
$^{-1}$, 215
$^\cup$, 215
$\circ$, 215
$\to$, 216
$\mapsto$, 216
$\rightsquigarrow$, 216
$''$, 216
$*$, 216
$[\![\ ]\!]$, 216
$^Y X$, 216
$\langle t \mid \varphi \rangle$, 216
$\approx$, 216
χ, 217

absolute, 134
absolute infinite, 2
absoluteness, 134
    for a theory, 136
abstraction term, 5, 27, 213
accumulation
    axiom of, 202
Ackermann, W., 10, 207

schema, 207
addition
  cardinal, 47
  for well-ordered sets, 68
  ordinal, 100
additively indecomposable, 104
aleph, 47
  function, 105
algebras
  regular open, 64
almost universal, 153
amorphous, 186
antichain, 121, 157
  maximal, 157
antisymmetric, 51
approximations, 24
Archimedean property, 91
arithmetic
  cardinal, 46
  language of, 76
  of the alephs, 106
  ordinal, 70, 100
assignment, 19
  satisfying, 130
  variant, 21
AtFml, 129
atomic formula, 15
atoms, 63
Aussonderungsaxiom, 29
axiom
  Martin's, 178
  multiplicative, 110
  null-set, 28
  of accumulation, 202
  of choice, 4, 31, 109, 164, 204
    class form, 195
    in L, 145
    independence of, 179
  of comprehension, 194
  of constructibility, 128
  of determinacy, 14, 124
  of extensionality, 27, 204
  of foundation, 30, 202, 205
  of infinity, 30, 73
  of replacement, 98
  of restriction, 202
  of separation, 29
  pair-set, 28, 159
  power-set, 29, 161, 217
  replacement, 29, 162, 195
  subset, 29, 195

sum-set, 28, 160
union, 28
weak power-set, 142
weak replacement, 142
axioms
  for inaccessible cardinals, 200
  logical, 18
  of extent, 33, 199
  Peano, 76
  using ranks, 202
  ZF, ZFC, 27

back and forth argument, 93
Banach, 43
basic definitions, 213
basis, 72
Bernays, P., 10, 153, 193
bi-implication, 16
Boole, G., 196
Boolean
  algebra, 58
  operations, 214
  value, 189
  valued models, 154, 189
bound
  upper, 54
bound variable, 16
bounded above, 54
brackets, 18
Brouwer, L. E. J., 210
Burali-Forti, C., 1
  paradox, 7, 95

c.a.c., 166
c.c.c., 166
Cantor, G., 1, 66, 193
  alephs, 106
  back and forth argument, 93
  construction of $\mathbb{R}$, 92
  diagonal method, 42
  normal form, 104
  paradox, 7, 43
  theorem, 42
Card, 47
cardinal
  addition, 47
  arithmetic, 46
    of the alephs, 106
  collapsed, 169
  comparability of, 113
  exponentiation, 47, 120, 127

INDEX 227

hyperinaccessible, 201
inaccessible
   axioms for, 200
   limit, 107, 200
   measurable, 200
   multiplication, 47
   number, 46
   preserved, 165
   strongly inaccessible, 200
   weakly inaccessible, 200
cartesian product, 215
   of family, 217
Cauchy sequence, 92
chain, 52
   descending, 66, 118
chain condition
   λ-, 170
   countable, 166
characteristic function, 49
choice
   axiom of, 4, 31, 109, 164, 204
      class form, 195
      in L, 145
      independence of, 179
   dependent, 10, 118
   function, 109, 188
   set, 4, 31
circular, 6
class form of the axiom of choice, 195
classes
   proper, 10, 193
   variables for, 193
closed formula, 16
closed unbounded sets, 122
closed upward, 156
closure
   universal, 18
club, 122
   filter, 122
cofinality, 119
cofinite, 59
Cohen, P., 4, 127
   real, 164
collapse
   transitive, 150
collapsed, 169
collapsing map, 150
combinatorial, 208
compactness theorem, 22
comparability of cardinals, 113
compatible, 155

complement, 59, 214
   relative, 62
complete
   first-order logic, 19
   lattice, 56
   subalgebras, 191
completeness theorem, 22
comprehension
   axiom of, 194
   predicative, 194
comprehension principle
   general, 5
   restricted, 29
   stratified, 204
condition
   λ-chain, 170
   countable antichain, 166
   countable chain, 166
   forcing, 155
cone, 64
conjunction, 16
connectives, 16
Connex, 94
conservative extension, 24, 196
consistency strength, 209
consistent, 14, 156
constant symbols, 128
constructibility
   axiom of, 128
constructible
   order, 141
   sets, 128
constructivist, 210
contains, 214
continuous, 122
continuum hypothesis, 45, 165
   generalized, 45, 128
converse, 52
   relation, 215
corners, 129
countable, 37, 38
   antichain condition, 166
   chain condition, 100
countable sets
   hereditarily, 108
course of values, 97
cub, 122
cumulative
   hierarchy, 103
   type structure, 8, 13
   types, 8

$\mathcal{D}$, 128, 132
decimal representations of reals, 93
Dedekind, R., 1
  finite set, 184
  infinite, 38
  recursion theorem, 77
deduction, 19
definability lemma, 158, 171
definable power-set, 128
definable Skolem hull, 152
definition
  by transfinite induction, 97
  recursive, 77
delta-system lemma, 166
  stronger versions, 170
dense, 155
denumerable, 38
dependent choices, 10, 118
descending chain, 66, 118
descriptive set theory, 125
determinacy
  axiom of, 14, 124
determined, 124
diagonal intersections, 123
diagonal method, 42
disjunction, 16
distributive, 57
diverge, 123
dom, 215
domain, 19, 215
dominance, 39
dual, 52

Easton, W., 127
Efml, 130
element, 3
elementary
  extension, 150
  number theory, 84
  substructure, 149
empty set, 213
  axiom, 28
enumerable, 38
equinumerous, 37
equivalence, 16
equivalent
  numerically, 37, 216
excluded middle, 210
expansion, 197
exponentiation
  cardinal, 47, 120, 127

  for well-ordered sets, 69
  ordinal, 101
extension, 63
  conservative, 24, 196
  elementary, 150
  generic, 154
extensional, 3
extensional elements, 191
extensionality
  axiom of, 3, 27, 204
  weak forms, 33
extent
  axioms of, 33, 199

families, 217
filter, 60
  club, 122
  maximal, 120
  prime, 61
  proper, 60
filter of subgroups
  normal, 179
finite, 37
  hereditarily, 33, 34
finite character, 116
finitism
  strict, 211
first-order logic, 13, 208
fixed points, 107
FM, 188
Fml, 130
FmlSeq, 130
Fn, 216
Fodor, G., 123
  theorem, 123
forcing, 127, 154
  conditions, 155
  iterated, 176
  language, 157
  models, 154
  product, 177
  relation, 155, 158
  strong, 171
  weak, 171
formalist, 210
formula, 15
  atomic, 15
  closed, 16
  well-formed, 16
foundation
  axiom of, 30, 202, 205

INDEX

Fraenkel, A. A., 10
Fraenkel–Mostowski models, 188
free variable, 16
Frege, F. L. G., 1, 204
Func, 216
function, 215
  characteristic, 49
  choice, 109, 188
functional, 29, 216
functions
  Skolem, 148
fuzzy sets, 4

games, 124
generality, 22
generalization, 19
generalized continuum hypothesis, 45, 128
  in L, 148
generated
  filter, 63
  ideal, 63
generic, 155
  extension, 154
  subset, 154
  ultrafilter, 190
good, 78, 98
greatest lower bound, 54
ground model, 154
grounded, 6
  sets, 10
groundedness, 6
  paradox of, 203
groundless, 6
Gödel, K., 4, 10, 127, 193
  theorems, 207
  constructible sets, 128
Gödel-set, 129

Hartogs
  aleph function, 105
Hausdorff, F., 114
hereditarily
  countable sets, 108, 218
  finite sets, 33, 34, 103
  symmetric, 180
Hilbert, D., 2, 211
hsym, 180
hyperinaccessible cardinals, 201
hypotheses, 19

ideal, 60
  generated by, 63
  maximal, 63, 120
  prime, 61
  proper, 60
Ifml, 130
implication
  bi-, 16
  material, 16
inaccessible
  cardinals, 200
  weakly, 200
incomparable, 52
incompatible, 155
incompleteness theorems, 209
inconsistent, 2
Ind, 72
indecomposable ordinal, 104
independence
  of the axiom of choice, 179
index-set, 217
indexed, 217
individuals, 4
induction, 67
  hypothesis, 72
  mathematical, 72
  step, 72
  transfinite, 97
inductive, 72
inference
  rules of, 19
infimum, 54
infinite, 37
  absolute, 2
  Dedekind, 38
infinite sequence, 217
infinity
  axiom of, 30, 73
infinity lemma, 121
initial
  ordinal, 105
  segment, 67
inner model, 145
Int, 72
integer, 72
  standard, 82
interpretation, 19
intersection, 214
  diagonal, 123
iterated forcing, 176
iteration, 77

Jech, T. J., 153
Jensen, R. B., 154, 205
join, 56

Kelly, J.L., 197
König, D.
  infinity lemma, 121
König, J.
  lemma, 120

L, 133
L-**good**, 144
λ-chain condition, 170
language
  forcing, 157
  of arithmetic, 76
last difference, 68, 69
lattice, 56
  complete, 56
  set, 60
least number principle, 67
least upper bound, 54
left Dedekind section, 90
level, 7
Levy, A., 10
lexicographic
  reverse, 68
limit
  cardinal, 107, 200
  ordinal, 95
limitation of size, 32
linear, 51
logic
  first-order, 13, 208
  second-order, 23
logical
  axioms, 18
  connectives, 16
  predicate, 15
logically valid, 21
lower bound, 54
  greatest, 54
LST, 13
  semantics of, 19
  syntax of, 14
Löwenheim–Skolem theorem, 198

map
  collapsing, 150
  order-preserving, 53
Martin's axiom, 178

material implication, 16
mathematical induction, 67, 72
maximal, 53
  antichain, 157
  filter, 120
  ideal, 63, 120
  principles, 114
measurable cardinal, 200
meet, 56
member, 4
meta-variables, 15
minimal, 53
Mirimanoff, D., 8, 93
MK, 197
model, 22
  Boolean-valued, 154, 189
  forcing, 154
  Fraenkel–Mostowski, 188
  ground, 154
  inner, 145
  natural, 201
  symmetric, 180
modular, 65
modus ponens, 19
Montague, R. M., 202
Morse, A.P., 197
multiplication
  cardinal, 47
multiplication for well-ordered sets, 68
multiplicative axiom, 110
multiplicatively indecomposable, 104
multisets, 4

name, 157
natural model, 199, 201
natural numbers, 71
  uniqueness of, 81
NBG, 193
negation, 16
Nfml, 129
nominalism, 211
non-circular, 6
non-logical, 15, 20
non-standard models, 83, 210
non-stationary, 122
non-trivial, 156
nondenumerable, 41
normal filter of subgroups, 179
normal form, 104
null-set axiom, 28
number

INDEX 231

cardinal, 46
natural, 71
rational, 85
real, 89
numerically equivalent, 37

object symbols, 15
objects, 4
od, 141
order relation, 51
order-complete, 55
order-isomorphism, 53
order-preserving maps, 53
order-types of well-orderings, 101
ordered
    pairs, 214
    set, 51
    triple, 215
ordering by last difference, 68
orderings, 51
orders, 7
ordinal, 66, 93
    arithmetic, 70, 100
    indecomposable, 104
    initial, 105
ordinal exponentiation, 101

$\mathcal{P}$, 214
$\mathbb{P}$, 214
PA, 76
pair-set
    axiom of, 28, 159
paradox
    Burali-Forti, 7, 95
    Cantor's, 7
    of groundedness, 6, 203
    of non-circular sets, 6
    Russell's, 5, 203
    Skolem, 198
    Tarski's, 199
parameters, 17
parse tree, 130
partial, 51
    choice principles, 117
Peano, G.
    axioms, 76
    system, 81
philosophy of mathematics, 207
platonist, 83, 210
poset, 51
positive and negative integers, 85

power-set, 214
    axiom, 29, 161, 217
    weak, 142
    definable, 128
pre-dense, 157
predicate symbols, 15
predicative, 194
predicative comprehension principle, 194
prenex form, 26
preserved, 165
prime
    filter, 61
    ideal, 61, 118
principal
    filter, 61
    ideal, 61
product
    cartesian, 215
    of family, 217
    forcing, 177
    relational, 215
    restricted, 68
product well-ordering, 68
projections, 132
proof, 19
proper classes, 10
proper ideals, 60

quantifier
    existential, 16
    universal, 16
Quine, W. van O., 4, 10, 204

ramified theory of types, 7
Ramsey, F.P., 9
ran, 215
range, 215
ranks, 9, 103
    variables for, 202
rational numbers, 85
real
    Cohen-generic, 165
    numbers, 89
        Cantor's construction, 92
        decimal representation, 93
realist, 83, 210
recursion
    course of values, 97
    over general well-founded relations, 103
    theorem, 45, 77

recursive definitions, 77
reductionism
  set-theoretic, 215
reflection principles, 138
reflexive, 51
regressive, 123
regular, 119, 200
regular open algebras, 64
regularity, 119
Rel, 215
relation, 215
  converse, 215
  forcing, 155, 158
  restriction of, 215
  symbols, 15
relational product, 215
relative complements, 62
relative consistency results, 46
relativized, 134, 205
replacement
  axiom of, 29, 98, 163
restricted comprehension principle, 29
restricted product, 68
restricted variables, 72
  for sets, 194
restriction
  axiom of, 202
restriction of relation, 215
reverse lexicographic, 68
Robinson, A., 210
root, 167
rules of inference, 19
Russell, B., 1
  paradox, 203

Sat, 132
satisfaction relation, 20
satisfying assignments, 130
SatSeq, 132
Schröder–Bernstein theorem, 43
scope, 16
Scott, D. S., 9, 202
second number class, 105
second-order logic, 23
section
  left Dedekind, 90
semantic consequence, 22
semantic paradoxes, 7
semantics, 19
sentence, 16
separation
  axiom of, 29
  separative, 177
  quotient, 177
sequence, 217
  Cauchy, 92
  infinite, 217
set algebra, 60
set lattice, 60
set-theoretic reductionism, 215
similar, 37, 53
similarity, 53
simple type structure, 8
singular, 119, 200
Skolem, T. A., 5
  functions, 148
  hull, 149
    definable, 152
  paradox, 198
Solovay, R. M., 127
sorts, 34
sound
  first-order logic, 19
sparse, 170
spec, 146
splits, 156
stabilizer subgroup, 180
standard integers, 82
standard model for ZF, 83
standard transitive structures, 130
stationary, 122
Stone representation theorem, 60
strategy, 124
stratified, 204
  comprehension principle, 204
strict, 51
strict finitism, 211
strong forcing, 171
strongly inaccessible cardinal, 200
strongly inductive, 197
structure, 19
  standard transitive, 130
sub-ordering, 52
sublattice, 56
subset, 214
  axiom of, 29
substructure
  elementary, 149
successor, 72
  ordinal, 95
sum of well-orderings, 68
sum-set

INDEX 233

axiom of, 28, 160
supertransitive, 206, 207
support, 69, 177
supremum, 54
sym, 180
symbols, 15
  constant, 128
symmetric, 179
  difference, 214
  hereditarily, 180
  model, 180
symmetry group, 180
syntax of LST, 14

Tarski, A.
  paradox, 199
Term, 129
term, 15, 157
  abstraction, 5, 27
theory, 22
total, 51
transfinite, 2
  induction, 97
    definition by, 97
transitive, 35, 51
  closure, 84
  collapse, 150
  set, 74
  structures, 130
trichotomy, 75
  for ordinals, 94
true in a structure, 20
truth lemma, 158, 171, 173
Tukey–Teichmüller lemma, 116
type structure
  cumulative, 8, 13
  ramified, 8
  simple, 8

ultrafilter, 118
  generic, 190
unbounded sets, 122
uncountable, 41
uniformized, 111
union, 214
  axiom of, 28
uniqueness
  of $\mathbb{R}$, 91
  of the natural numbers, 81
unit, 58
universal

closure, 18
interpretation, 22
universally valid, 21
universe, 19
unnaturalness, 87
unstratified, 204
upper bound, 54
  least, 54
urelements, 4, 188

$V = L$, 128
$V^{\mathbb{B}}$, 189
valid, 21
value
  Boolean, 189
variable
  symbols, 15
variables
  bound, 16
  for ranks, 202
  free, 16
  restricted, 72
  for sets, 194
variables for classes, 193
variant assignment, 21
vicious circle principle, 7, 194
Vlength, 131
VNB, 128, 193
VNBG, 193
von Neumann, J., 2, 9, 71, 93, 193, 205

weak forcing, 171
weak power-set axiom, 142
weak replacement axiom, 142
weakly inaccessible, 200
well-formed-formulas, 16
well-founded relations
  recursion over, 103
well-ordered, 66
well-ordering theorem, 112
  Zermelo's proof, 113
well-orderings
  order-types of, 101
Whitehead, A. N., 7
winning strategy, 124
wo, 112

yields, 19

Zermelo, E., 2
Zermelo–Fraenkel axioms, 27

zero, 58
ZF, 27
ZFA, 188
ZFC, 27
Zorn's lemma, 114